Green Energy and Technology

Climate change, environmental impact and the limited natural resources urge scientific research and novel technical solutions. The monograph series Green Energy and Technology serves as a publishing platform for scientific and technological approaches to "green"—i.e. environmentally friendly and sustainable—technologies. While a focus lies on energy and power supply, it also covers "green" solutions in industrial engineering and engineering design. Green Energy and Technology addresses researchers, advanced students, technical consultants as well as decision makers in industries and politics. Hence, the level of presentation spans from instructional to highly technical. **Indexed in Scopus**.

More information about this series at http://www.springer.com/series/8059

Pere Mir-Artigues · Pablo del Río · Natàlia Caldés

The Economics and Policy of Concentrating Solar Power Generation

Springer

Pere Mir-Artigues
Energy Sustainability Research Group
University of Barcelona
Barcelona, Spain

Natàlia Caldés
Energy Systems Analysis Unit
Centro de Investigaciones Energéticas
Medioambientales y Tecnológicas
(CIEMAT)
Madrid, Spain

Pablo del Río
Institute for Public Policies and Goods
Consejo Superior de Investigaciones
Científicas (CSIC)
Madrid, Spain

ISSN 1865-3529 ISSN 1865-3537 (electronic)
Green Energy and Technology
ISBN 978-3-030-11937-9 ISBN 978-3-030-11938-6 (eBook)
https://doi.org/10.1007/978-3-030-11938-6

This Springer imprint is published by the registered company Springer Nature Switzerland AG
The registered company address is: Gewerbestrasse 11, 6330 Cham, Switzerland

Foreword

The climate change caused by the massive use of fossil fuels has become more and more evident during this century, and a more sustainable energy market is therefore needed to avoid a global warming above 2°C, which would put the mankind in a dramatic situation. In this context, solar energy is seen as one of the best options to reduce the consumption of fossil fuels, not only in the electricity market but also in the industrial sector in general because heat represents three quarters of the worldwide industries' energy demand, and 90% of this heat (79.5 EJ) is currently supplied by fossil fuels.

Concentrating solar thermal (CST) systems are internationally recognized as a key element among the different options to use solar energy and thus, reduce the greenhouse gas (GHG) emissions. The facilities using CST systems to produce electricity, the so-called solar thermal electricity (STE) plants, or more traditionally concentrating solar power (CSP) plants, are showing a good reliability and high dispatchability. With only slightly more than 5 GW of total installed power, STE plants have already achieved a significant cost reduction and they still have a significant potential to reduce costs further. *Cost* is a major issue when talking about the use of renewable energy systems to produce electricity, because unfortunately this is usually the only item taken into consideration by policymakers, investors, and grid operators when comparing the conventional fossil-fueled technologies with modern clean renewable options. The economic added value of dispatchability and "spinning reserve" for the grid are not included in most of cost analysis, and it is therefore critical to analyze STE systems with alternate economic metrics, because the so-called Levelized-Cost-of-Electricity (LCOE) is not a sufficient indicator of its value for the electrical system. This fact is a significant barrier for STE plants because their great socio-, techno-, and economic benefits are not taken into account when performing this comparison.

This book clearly depicts how arduous the process of economic optimization of STE plants is because of the many technical and economic variables involved. The detailed cost analysis of STE plants is quite complex because of the many parameters and boundary conditions that must be taken into account. This is the main reason why the availability of detailed information is essential to guarantee the usefulness of the results obtained from a STE plants' feasibility study, which must include not only economic parameters, but also many other parameters and considerations. This book provides the reader with this type of data and information, which will help you not only to get acquainted with the complexity associated with the economics of STE plants, but also to get a comprehensive knowledge of the current worldwide situation and trends of this technology with detailed information about the national plans for STE plants in those countries currently promoting these systems.

The three authors of this book, Dr. Pere Mir-Artigues, Dr. Pablo del Río, and Dr. Natàlia Caldés, have an excellent background and experience in the subject of this book. They have performed an extensive research on renewable energy systems, covering lifecycle assessments and analysis of support schemes for renewable electricity, the interactions of climate and renewable energy policies, and on the drivers to eco-innovation in energy, industry, and the energy and transport sectors. During their professional life, they have acquired a valuable overview of renewable energy systems and their socio-, techno-, and economic issues. This book can be considered a melting pot of their valuable experience applied to STE plants.

When reading this book, I quickly became aware of the huge effort devoted by the authors to consult and look into many journal articles, items from newspapers, reports from relevant associations and institutions, publications exclusively dedicated to STE sector, and relevant energy journals, thus obtaining and putting to the disposal of the reader a very comprehensive information about not only cost analysis and technological trends, but also about the drivers and barriers of this technology. Another important contribution of this book is the complete analysis performed by the authors of the social value of solar thermal electricity in a changing electricity system by identifying its main role in the different stages of the structural change of the electricity sector, thus determining the value of solar thermal electricity, which is understood as its contribution to the success of the energy transition, that is, to the evolution toward an electricity system with a dominant role of renewable energy sources. The authors also give interesting ideas and a vision of the possible management and key pillars for the sustainability of a decentralized and basically renewable electricity sector.

In summary, this book will provide you with a quite complete overview of the past, present, and possible future of solar thermal electricity, which is considered a key element to evolve toward a green electricity market due to its socioeconomic benefits and complementarity with less dispatchable renewable energy options.

Almería, Spain

Eduardo Zarza Moya
CIEMAT-Plataforma Solar de Almería

Contents

Main Acronyms, Symbols and Technical Units

AC	Alternating current
BE	Balancing effect
CSP	Concentrating solar power
DC	Direct current
DNI	Direct normal irradiance
E_{th}	Thermal energy required by the process
E_{th}^{F}	Nominal thermal power of the solar field
EJ	Exajoule (1018 joules)
EcV	Economic value
EnV	Environmental value
EPC	Engineering, procurement, and construction
FIPs	Feed-in premiums
FITs	Feed-in tariffs (preferential tariffs)
GHG	Greenhouse gases
GrC	Grid-related costs
HTF	Heat transfer fluid
HVAC	High-voltage alternating current
HVDC	High-voltage direct current
L	Capacity factor
LCOE	Levelized-Cost-of-Electricity
MENA	Middle East and North Africa
MIV	Market integration value
MOE	Merit order effect
MT	Metric tonnes
O&M	Operations and maintenance
PPA	Purchase power agreement
RES-E	Renewable energy sources-electricity
RD&D	Research, development, and demonstration
SM	Solar multiple
SV_{CSP}	Social value of concentrating solar power

SO	System operator
TES	Thermal energy storage
TSO	Transmission system operator
T&D	Transmission and distribution
WACC	Weighted average cost of capital
c	Renewable generation cost
e, e_t	Retail electricity price
i	Interest rate (and discount rate when applicable)
H	Hours of the year (8760)
m_t	Maintenance expenditures in the year t (monetary units)
p	Price by MWh received by generators (could be a preferential price)
q	Annual generation MWh of AC
q^*	Net annual generation MWh of AC
w	Wholesale electricity price or tariff, as indicated
Λ	Installed capacity (kW or MW)
Λ_{th}	Installed thermal capacity (kW or MW)
λ	Premium per kWh or MWh
kV	Kilovolts
kW, kWh	Kilowatt, kilowatt hour
MW, MWh	Megawatt, megawatt hour
GW, GWh	Gigawatt, gigawatt hour
TW, TWh	Terawatt, terawatt hour
kW_{th}, MW_{th}	Kilowatt thermal, megawatt thermal

Chapter 1
Introduction

Solar thermal electricity generation, or concentrating solar power (CSP), is the production of electricity using direct solar irradiation as the primary source of energy. The amount of solar irradiation directly coming from the sun, wherever is its position in the sky, is called Direct Normal Irradiance (DNI). The first step to obtain electricity is to concentrate the solar rays to heat a fluid to a temperature which is sufficiently high to produce steam after its transit through a heat exchanger. Next, depending on the pressure and the temperature reached by the steam, a specific type of turbine connected to an electricity generator is activated. Given the high capacity of the power plant (in the order of MW), the electricity is evacuated through a transmission grid. Concentrating solar plants are, thus, a chain of energy conversions: In order to obtain electricity, a given working fluid accumulates the thermal energy contained in the direct solar irradiation which has been concentrated by the appropriate collectors. Then, the heat of the fluid transforms water into superheated steam at a given pressure in a heat exchanger. The steam moves the blade of a turbine whose axis is connected to a rotor (or inductor) which rotates at a high speed within an attached coil to a core of ferromagnetic material (or stator). In this manner, the mechanical energy of a turbine ends up being transformed into alternating current [10: 172]. Or, put in simpler terms, a primary source of thermal energy, in this case renewable, is converted into electricity after having transfered and converted the heat into mechanical energy. Therefore, solar thermoelectric generation can be classified according to two different perspectives:

- As a renewable source which exploits solar irradiation, a feature shared with solar photovoltaic generation, although this is the only common feature.
- As a thermal plant which, instead of using coal, oil, gas, or enriched uranium as a primary source of heat, takes advantage of the rays which are directly coming from the sun.

P. Mir-Artigues et al., *The Economics and Policy of Concentrating Solar Power Generation*, Green Energy and Technology,
https://doi.org/10.1007/978-3-030-11938-6_1

Most CSP plants are made up of the following three elements [6: 51–71, 8: 1–2]:

- The solar field, whose mission is to capture, reflect, and concentrate the direct solar irradiation. There are three main methods: an ordered set of parabolic trough collectors (or, in a variant, linear ones) in whose focal axis the solar rays are concentrated, an extension of mirrors (or heliostats) which reflect those to an upper focal point, or a set of mirrors which make up a paraboloidal dish where rays are directed toward the focus. In all these cases, the captured heat is accumulated by a heat transfer fluid (HTF) or working fluid which, in the trough collectors and heliostats field, goes through a circuit of considerable length. In a heat exchanger, the thermal energy of the HTF is transferred to water, which is then turned into steam to the appropriate temperature and pressure to activate the corresponding power cycle. It should be noted that the HTF circuit contains auxiliary devices to keep its temperature above a given level. If the working fluid was only pressurized steam, the thermal process would then be more direct. However, this design must deal with technical challenges which are difficult to solve [3, 7: 37]. The solar field, which is the distinctive element of solar thermoelectric generation, requires a large area: between 20 and 35 thousand m^2 per kW in the case of parabolic trough technology, between 12 and 35 thousand m^2 per kW for solar towers, and around 35 thousand m^2 per kW for the dish/Stirling engine. Therefore, around 2 hectares per MW of capacity [4: 10] or between 15 and 60 MW/km^2 are required, depending on the plant design and the storage capacity [2: 10–56]. Although the land covered by the solar field can be used for animal grazing, such option cannot be recommended due to security reasons.
- Thermal energy storage (TES), or tanks where part of the heat being carried by the HTF, is accumulated. Given the fact that solar irradiation is interrupted from dusk to dawn, two tanks are available to enable the electricity generation during the night hours: the hot tank, where the heat captured by the solar field is stored for its use during the night, and the cold tank, where the working fluid which yielded heat before returning to the solar field is accumulated. Thermal oil (or, more commonly, molten salts) used as HTF can be stored. Molten salts are cheaper and are neither toxic nor inflammable [7: 36]. Mass storage using concrete, or heat accumulation using the so-called phase-change materials, has also been proposed [10: 177]. For the latter, a heat exchanger between the working fluid and the stored heat is required. Although the heat storage capacity is measured in thermal units, the most common way to express the storage capacity is in terms of the additional operating hours of the plant as they are both considered equivalent due to the small TES losses. Given the considerable size of the tank, often above 20 m of diameter and height, the required steel and the refractory materials account for more than 90% of its total costs.
- The power block is a system made up of a turbine and an electromagnetic induction generator. Its nominal power is measured in MW. A thermodynamic cycle takes place in the power block, and this is why it is also called power cycle. There are several types of steam turbines. The most common in thermo-solar plants are those operating according to a Rankine cycle (with variants such as overheating,

regeneration, and reheating). All of them use steam at moderate or high temperatures.[1] All these technologies are very well known, since they have been used in conventional thermal plants for decades. Since efficiency increases with temperature, the circuit feeding the power block can have various designs. For example, steam may be generated by a HTF/water interchanger which moves a first body of the turbine, with the steam being reheated again with a fluid coming from the solar field. This is later directed (at a lower pressure) to a second body of the turbine. Hybridization is another possibility: The steam obtained after the HTF/water interchanger increases the temperature in a boiler which is heated by natural gas or any other fossil fuel or biomass. In hybridized plants, the type and pattern of fuel which may be burnt every hour, or only during those hours in which there is no solar irradiation as a complement to heat storage, depend on the regulation and the evolution of their prices.[2] Finally, the set of the power block includes a cooling system,[3] equipment which increases the voltage of the electricity before it feeds the transmission grid, etc.

As a general rule, thermo-solar plants transform 15–20% of the incident solar energy into electricity which is fed into the grid. The greatest losses take place in the solar field and the heat transfer systems (~60% of the total solar heat captured is not converted into superheated steam), whereas the power block loses an additional 25%.

The best locations for CSP plants are between 20° and 35° of north and south latitude. These are two wide strips of the Earth's surface, where the greatest amount of solar thermal electricity can be generated. The desert and arid regions cover 25.4 million km^2, half of which are located in Africa (Sahara, Kalahari, Namib, and Ogaden deserts) and the rest are located in Australia (Gibson, Great Sandy, etc.), Middle East (deserts in the Arabian peninsula, Iran, South of Afghanistan, and Balochistan), Western China (Ordos, Lop, Mu Us, etc., plus the Gobi desert shared with Mongolia) and Atacama (Chile), together with the deserts in the Peruvian coast, the Bolivian high plateau as well as Western Argentina, USA (states of Arizona, California, New Mexico, and Nevada) and Mexico (deserts in Baja California, Sonora, etc.). In all these regions, the DNI is above 2000 kWh/m^2/year. In some of these places, the DNI may even be twice this level.[4] However, what is important is that the DNI is above 2000 kWh/m^2/h when plants are in operation. This is a huge territory where the potential for electricity generation is estimated at 3 million TWh/year, an amount

[1]It is also possible to generate solar thermal electricity using a gas turbine, in which the operating principle is the Brayton thermodynamic cycle. However, this is not a commercial option yet.

[2]Hybrid plants need a receiver installation which is able to use natural gas, biomass, etc.

[3]There is the dry and the wet cooling system. Both have similar costs, although the annual electricity generation of the former is between 3% and 6% lower. The warmer the environment, the more likely that the performance of the wet-cooled system regarding the condensation of the steam exhaust exceeds the performance of the dry-cooled system [2: 10–23].

[4]Desert and arid regions which are less suitable for CSP include those where there are, even if unlikely, rainy events, the phenomenon of fog desert and fields of constantly moving dunes, or have deep slopes.

which is many orders of magnitude higher than the 18 thousand TWh/year which are currently consumed in the world [5: 7, 9: 1–3].

Additional important applications of CSP which are worth highlighting include the generation of hydrogen and water desalination. The latter shows good prospects, since it is a suitable complement: Those places with the highest solar irradiation areas in the world (deserts and arid zones) are often those with scarcer water resources, which are insufficient to meet the needs of the population (residents and tourists), farm production, and industrial activities. The capacity of the solar field to produce high-temperature and high-pressure steam has also raised the interest for many other uses, including the purely industrial uses, the improvement of the efficiency of coal power plants and the thermal enhanced oil recovery [1] as well as pumping irrigation water with a steam engine. There is not any power block in the above-mentioned applications and, thus, electricity is not generated.

This book provides an introduction to the economic analysis of concentrating solar power generation, which involves the design of an appropriate analytical framework and the development of a given thematic repertoire. It is a handbook which includes the following main sections:

- An overview of the main technical features of CSP plants.
- A description of the historical and recent facts.
- An economic analysis of CSP generation.
- A scrutiny of the public support policies for CSP.

The authors acknowledge the help or/and comments provided by Félix Fernández Álvarez (CSIC), Christoph Patrick Kiefer (CSIC), Johan Lilliestam (REPG-ETH Zurich), Loreto Valenzuela (CIEMAT-PSA), and Eduardo Zarza (CIEMAT-PSA). Obviously, the authors are the only ones responsible for any mistakes or omissions that the text could still contain.

References

1. Chaar M et al (2015) Economics of steam generation for thermal enhanced oil recovery. Oil Gas Facil 4(6):42–50
2. Drury E et al (2012) Solar energy technologies. In: Hand MM et al (eds) Renewable electricity futures study. Volume 2: renewable electricity generation and storage technologies. Report NREL/TP-6A20-52409. National Renewable Energy Laboratory (NREL), Golden, CO, pp 10-1–10-60 (Chapter 10). Available at http://www.nrel.gov/analysis/re_futures. Accessed July 2014
3. Hirsch T et al (2013) Advancements in the field of direct steam generation in linear solar concentrators—a review. Heat Transfer Eng 35(3):258–271
4. IEA (2010) Technology roadmap. Concentrating solar power. OECD/IEA, Paris. Available at http://www.iea.org. Accessed Jan 2011
5. IRENA (2012) Concentrating solar power. Renewable energy technologies: cost analysis series. Volume 1: power sector, vol 275. Available at www.irena.org/publications. Accessed May 2014
6. MIT (2015) The future of solar energy. An interdisciplinary MIT study. Massachusetts Institute of Technology, Cambridge. Available at http://mitei.mit.edu. Accessed May 2015

7. Palenzuela P et al (2015) Concentrating solar power and desalination plants. Engineering and economics of coupling multi-effect distillation and solar plants. Springer, Berlin
8. Sioshansi R, Denholm P (2010) The value of concentrating solar power and thermal energy storage. TP-6A2-45833. NREL, Golden, CO. Available at http://www.nrel.gov. Accessed June 2012
9. Trieb F et al (2009) Global potential of concentrating solar power. In: SolarPaces conference, Berlin. Available at http://www.desertec.org. Accessed Apr 2010
10. Weinrebe G, Ortmanns W (2007) Principles. In: Kaltschmitt M et al (eds) Renewable energy. Technology, economics and environment. Springer, Heidelberg, pp 171–181

Chapter 2
Concentrating Solar Power Technologies: Solar Field Types and Additional Systems

Different CSP generation technologies can be distinguished depending on the type of collector's optics and solar receiver. In particular, they differ according to the geometrical shape and spatial placement of the mirrors, which determine the degree of concentration of DNI in the solar collector. This variable affects the temperature reached by the working fluid which circulates through the receiver and, hence, the type of turbine to be used and the efficiency of the transformation process between thermal energy and electricity [8, 24]. If the steam reaches a temperature around 400 °C, then a Rankine cycle turbine has an efficiency below 35%, whereas the efficiency is 10% higher at 600 °C [13: 14]. At higher temperatures, an even more efficient gas turbine can be used.

The following pages provide details on the technical and economic features of the main solar thermal technologies, with a particular reference to the solar field, i.e., the field of parabolic trough collectors, the linear Fresnel collector, the solar tower, and the paraboloidal dish coupled to a Stirling engine. The numerous variants of these four technical modalities, which are considered canonical by the experts, have not been addressed. At the end of the chapter, we refer to the technologies to store heat and the role of CSP plants in water desalination.

2.1 Parabolic Trough Collector

The most mature technology and, thus, the one which has most often been used is the solar field made up of rows of parabolic trough collectors. A synthetic oil circulates throughout its linear focus and, boosted by pumps, it gradually increases its temperature. At the end of the circuit, the HTF is directed to a heat exchanger where water is transformed into steam at sufficiently high temperatures and pressure in order to activate a power cycle.

P. Mir-Artigues et al., *The Economics and Policy of Concentrating Solar Power Generation*, Green Energy and Technology,
https://doi.org/10.1007/978-3-030-11938-6_2

The collectors of the solar field are the first aspect that should be addressed. They are linear parabolic-shaped reflectors of about 100 or 150 m of length (one collector unit or solar collector assembly can include up to 12 modules of 12 m of length) and about 5 m of width. The modules are tied forming loops composed of two parallel rows of 600 m long in the standard configuration. The mirrors are about 4/5 mm thick and do not contain iron, an element which leads to the slightly green tone of common glass, which absorbs part of the incident light and reduces the efficiency of the equipment. The removal of iron leads to mirrors which reflect light by an additional 20%, but this makes their manufacturing more expensive. The face directed to the sun is a silver-coated glass mirror. With the aim to protect mirrors against dust accumulation and corrosion, a protective copper layer and three epoxy varnishes are later applied. This leads to high levels of reflectivity: an average of 93.5% [12: 24].

The modules are not single mirrors,[1] but they are the outcome of assembling small-size units of rectangular shape with a slight curvature. Their dimensions are approximately 1.57 × 1.4 m. For example, the so-called Eurotrough contains 28 mirror facets per module. Obviously, increases in its size will entail fewer pieces per module, although its manufacturing becomes more expensive. Assembling the different mirror facets results in a perfect dish. The mirrors are later hooked to a bearing structure through small ceramic plates which host the fixing elements. Those supporting structures, made up of galvanized steel and lately made of aluminum, can have different frames. They are fixed to the floor through pylons on concrete footings. Since the collector rows cannot shade each other, separating them three times their width is the rule being followed [29: 198].

Each collector has a control board and a hydraulic group. They also have wind sensors, fluid temperature sensors, solar irradiation sensors, etc. Among the control systems, we should mention some safety mechanisms in the event that the pumps moving the working fluid were interrupted. If this happened, the temperature of the fluid would rise without control, endangering the integrity of the installation. The collectors move hydraulically through electric engines. However, they have hydraulic accumulators just in case they have to be moved in the absence of electricity supply. These devices also avoid turning on the engines again and again, since the collectors move every 4 or 5 min. At night, the collectors remain on a standby position, i.e., with the reflecting face directed to the floor. This position offers the least resistance to wind and facilitates O&M activities.

The orientation of collectors is an important issue. Two options exist: They can be disposed in a north/south direction, oscillating daily from east to west. Alternatively, they can extend their axis in an east/west direction and move them according to the height of the sun in the sky, which changes according to the seasons. Despite the seasonal differences in intensity, the first configuration is the most common one as it allows capturing more irradiation throughout the year. This is an irrelevant problem in the tropics, although not in higher latitudes. The second orientation has the advantage

[1] The manufacturing cost of a single parabolic mirror with several meters width would be prohibitive, and its installation would be challenging.

of a lower variation of the incident angle of the direct solar irradiation throughout the year. Thus, in the latter case, the captured irradiation is more stable, although lower in annual terms [22: 93–94].

Solar tracking is assured by devices which periodically measure the irradiation and move the mirrors accordingly. At present, assuming that the longitude and the latitude of the plants are well known, the collectors can be moved following the apparent position of the sun by using the pertinent algorithms.

The receiver tube of solar rays is located in the linear focus which makes up the parabolic reflecting surface (sustained at about 1.7 m). This is a component made up of two concentric tubes: The fluid circulates in the inner tube, made up of stainless steel and coated with a high selective coating, whereas the external tube is made up of a very transparent glass. It is manufactured with highly pure sands which have been subjected to about 1650 °C in order for the material to be ductile and have a tubular shape. When in operation, the inner tube reaches a temperature of maximum 450 °C and absorbs more than 95% of the incident energy. In reality, the set behaves as a thermo, whose vacuum is controlled by chemical sponges [22: 91–92].

The manufacturing of the double collector tube is a very sophisticated process: It includes glass-to-metal joints, and it has to be strictly sealed to preserve the vacuum and must be capable to resist a high-temperature gradient [12: 45–46, 23: 47–48]. For this reason, the few firms which are able to manufacture them keep the process in secret. There are rotary unions and/or hoses in the extreme of the collector links, which connect the collectors with the rest of the network of pipes of the solar field. These connections operate under pressure, subject to high temperatures, and must be flexible. Since they are one of the most vulnerable parts of solar thermoelectric plants, engineers aim to reduce the number of such unions, guarantee their functionality and reliability (in order to avoid voltage drops), and reduce their cost [23: 18].

The spatial configuration of the collectors in the solar field has three possible alternatives, as shown in Fig. 2.1 (adapted from Günther et al. [12]: 65–68):

- Direct feeding: A simple disposition which consists of a conduit which returns from the power block and distributes the fluid among the rows of collectors. The problem of this configuration is the control of the important differences in pressure between the inflows to the different rows and the outflows.
- Indirect feeding: The HTF enters in the rows as a balanced flow, although this entails a longer conduit and a greater effort to maintain the pressure.
- Central feeding: The rows face each other so that the inflows and outflows are both directed toward the center. In this case, valves which lead to homogenous flows of the rows are required, although the length of the conduits is shorter.

There are many variants of a parabolic trough plant. For example, pilot plants have been built which directly heat water/steam to feed a turbine generator or direct steam generation (DSG) process. Although this is a less complex design (with a single circuit and HTF), the control of pressure and temperature of the steam faces considerable technical challenges [12: 59–61, 23: 65]. At the end of the last century, the Plataforma Solar de Almería investigated the real conditions of operation of a 5 MW plant of this type [8, 24].

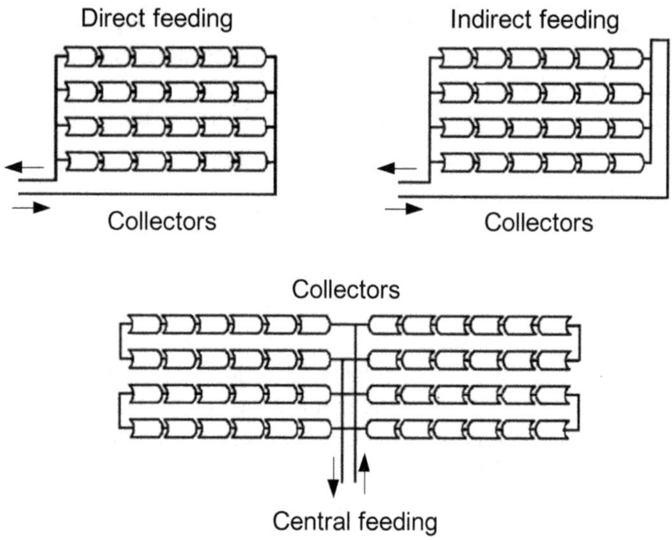

Fig. 2.1 Parabolic trough rows' layout

The improvements which can be incorporated into the collector and mirrors are diverse, but they all face cumbersome trade-offs, as shown by the following cases [1: 13–14, 25]:

- Increasing the length of the rows allows reducing the unions and connections of the solar field, but requires an increase of the capacity of the recirculation pumps.
- The thinner is the mirror, the more reflective but also the more fragile it is.
- If the reflecting face is only covered with aluminum foil, the cost is lower, but also the reflectivity (which falls below 90%).
- The mirror made up of polymer is the cheapest, but its durability is insufficient.

Therefore, the aim is to have the cheapest as possible reflective surface, without compromising its reflectivity.

Regarding the HTF, the so-called Therminol VP-1 is currently the most commonly used compound.[2] Its maximum bulk temperature is 400 °C, which is below the 500 °C at which absorber tubes can operate. Despite its low viscosity, the displacement of the tens of tons of HTF over the solar field requires the use of several pumps. Fortunately, its crystallizing point is about 12 °C, which only requires having reheating devices in latitudes where the temperature goes down significantly. Other oils can reach similar maximum temperatures, such as Syltherm 800, but their cost is much higher and their auto-ignition temperature is lower. Synthetic thermal

[2]This is a toxic mix of two hydrocarbons, the biphenyl, and the diphenyl oxide. It is also flammable, with self-ignition above 600 °C (see http://www.therminol.com). At the end of the last decade, its price was slightly above €10/kg.

oil has to be periodically replaced and properly recycled because of degradation due to age [12: 57–58, 22: 98].

The journey over the solar field, at a speed between 2 and 3.5 m/s, gradually increases the temperature of the HTF: Under regular functioning conditions, one hundred Celsius degrees is the difference between the inflow temperature (around 295 °C) and the outflow temperature (around 395 °C). This implies that the working fluid increases by about 2° for each collector module. It should be added that, when leaving the solar field, the working fluid has to be 10/15 °C above the working temperature of the steam which moves the power block [22: 97]. The circuit of the solar field can be emptied in its lowest point.

There are three operation situations regarding the circulation of HTF:

- Solar irradiation is very low or null, and thus, the oil cools down when going through the solar field. In order to reach the freezing point, the oil stops circulating in the solar field and is heated with gas or other fuels.
- Solar irradiation is still insufficient for the HTF to reach 400 °C. The oil will circulate over the solar field, as long as the energy captured offsets the energy consumed by the pumps which boost it. A gas boiler or a boiler with other fuels or a heat exchange with the heat being stored in the TES will then provide the necessary energy.
- Solar irradiation is strong and persistent, and thus, the oil is maintained for a long time at very high temperatures. Since this can degrade it, some collectors can be partially defocused.

Therefore, to make the most of solar irradiation, the oil has to enter the solar field at a minimum temperature because, otherwise, there is a risk that the HTF does not sufficiently increase. However, if it usually too hot, then it is likely that the size of the solar field was not properly calculated.

There are two separated circuits with different HTF in parabolic trough plants: the primary one, which extends over the solar field according to a studied network reaching the heat exchanger, and the secondary one, whose working fluid is steam and connects the latter with the turbine which is attached to an electricity generator. Both require electric pumps to boost the fluids, and they have a plethora of valves, sensors, etc. As expected and in both cases, the efficient functioning of the plant requires its alignment with strict pressure and temperature parameters.

Much research has been devoted to substitute synthetic oil as HTF. At present, molten salts represent the best alternative. They are a mix of sodium and potassium nitrates which is neither toxic nor inflammable. This substance is liquid above 238 °C, and it is able to absorb large quantities of heat since it reaches 600 °C. Notwithstanding, since it is solid at ambient temperatures, the molten salts have to be melted through electric resistances in small deposits distributed in the solar field. Only in a low viscosity state, it can be inserted into the pipes. It is necessary to avoid their solidification since they would block all the circuits, leading to a considerable problem. The need to maintain them at high temperatures at night explains their use in plants with TES and/or hybridization. Steam is another possible working fluid. The advantage is that the heat exchanger is removed, but the drawback is the need to

Table 2.1 HTF working temperatures

HTF	Temperature range (°C)
Synthetic oil	13/400
Organic (siliconic oil)	−40/400
Steam	0/>500
Molten salts	238/600
Ionic liquids	−75/416
Air	>500

work at high pressures. This is a stringent requirement for the receiver tubes. Finally, it should be mentioned that many other fluids, whose working temperatures appear in Table 2.1, have been considered.

As mentioned above, with the aim to maintain electricity generation active for as long as possible, parabolic trough plants can be hybrid (additional consumption of gas, biomass, etc.) or deposits to store heat can be added to them (see below). In addition, this technology (together with the Linear Fresnel collectors) may also be used to obtain steam at high temperatures for its use it in industrial activities, oil recovery, etc. In this case, the power block does not exit and the solar field systems are identical.

In parabolic trough plants, the electricity/solar conversion is between 15 and 20%, and it is expected to reach 18% in a few years [22: 38]. However, it should be mentioned that this low value is mostly due to the poor performance of the turbine as a result of the low operating temperature, as well as the losses due to the displacement of the fluid, given that the absorption tubes as well as the heat transfer process and the electricity generator have much higher efficiencies [29: 197]. Water requirements for the power block are estimated to be between 2.8 and 3.5 m^3/MWh [4: 8, 6: 10–46, 20: 49].[3] The cleaning of the collectors depends on the weather circumstances (rain, wind, etc.), but they would be 9–10 times lower per MWh. Finally, lifecycle GHG emissions of trough systems show the following values (minimum, 25th percentile, median, 75th percentile and maximum, respectively): 9, 14, 19, 26, and 49 g CO_2 eq/kWh (Edenhofer et al. [7: 372 and 982] and Drury et al. [6: 10–47]. See also Mai et al. [17]).

2.2 Solar Tower

A central receiver system or solar tower is a very different technology. In this case, a field is made up of hundreds or thousands of flat reflectors of the direct sun rays, called heliostats. These are subjected to a mobile structure which keeps them above the ground. The heliostats, as it is also the case with the concentrating photovoltaic modules, follow the solar transit and reflect the rays to the upper part of the tower,

[3] In case of dry cooling, the water consumption rate is around 0.25 m^3/MWh.

where a receiver point is located. This device focuses the solar rays from 200 to 1000 times, acting also as heat exchanger. Also, in this case, a HTF goes across the receiver where it is heated to very high temperatures (hundreds of °C). Then, this thermal energy is used to produce steam and move a turbine/generator. A solar tower plant configuration can also be a hybrid one.

The heliostats are flat glasses, originally of about ≤50 m^2 and later ≥120 m^2, which follow the solar transit in every moment. In fact, the heliostats are not a single piece, but are made up of multiple square segments to facilitate their assembling. Their reflectivity is above 95%, and an automatic mechanism ensures the tracking of the sun. This movement depends on the latitude and the season (the height of the sun above the horizon). The structure which supports the heliostats, often disposed in pairs, bears a weight of about 2 metric tons, and it must resist the winds gusts without moving. A minimal defocus of the heliostats will drastically reduce the energy performance of the system. An overcast sky is also a problem: Sudden changes in the temperature of the receiver may lead to the instability of the working fluid, especially if this is water/steam. On the other hand, all the previously mentioned remarks regarding the width, reflectivity, and composition of mirrors of the parabolic trough rows also apply to this case.

The greater the size of the heliostats, the lower is the size of the solar field and, thus, the required number of tracking systems. However, the robustness of the supporting structures has to be greater, especially because the wind may defocus them. To overcome this technical challenge, smaller heliostats are sometimes preferred (from 1 to 7 m^2), which allows saving on foundations and costs of installation of the solar field (from 1 to 7 m^2). This configuration leads to fewer shadows and simplified support requirements. However, the costs of tracking, control, and maintenance (basically, cleaning) increase. There does not seem to be a definitive design which, taking into account all the trade-offs, is unambiguously the most economically optimum option.

The spatial layout of the heliostats in the solar field is a relevant issue: One option is the radially staggered layout, and the other is the phyllotaxis spiral layout. This second arrangement allows fitting the maximum number of heliostats per space unit. This layout, which is identical to the sunflower seeds display, requires a comparatively smaller surface and leads to a higher optical efficiency [2: 21].

In the beginning of the solar tower technology, the cost of heliostats was high (>$250/$m^2$). However, it was strongly reduced by large-scale manufacturing, because the cost depends on annual production rates. For example, if around 50,000 heliostats are manufactured per year, then costs fall fivefold.

Towers usually have 100 m height. In their upper part, receivers of an appropriate geometry transform solar irradiation from the heliostats into thermal energy. The receiver design has been changing over the years. The traditional option was the tubular receiver: There was a coil behind the structure which captures the sun rays through which the working fluid circulated at a temperature below 500 °C [23: 74–75]. Later, the system which creates a convective flow of hot air which goes through a porous body (a metallic mesh or a ceramic surface), located in a special cavity, became the best procedure in terms of energy performance. This is the open volumetric air receiver, which is able to reach very high temperatures without energy losses. Behind this cavity, the circulating air is heated to between 500 and 800 °C, and it is able to produce steam at 480–540 °C in a Rankine cycle. If the air is pressurized behind a quartz window and the receiver is a ceramic one, >1000 °C can be reached, which opens the door to the use of turbines such as the ones used in natural gas plants. In this case, the performance of the plant could be as high as 50% [22: 39–40]. All in all, finding materials which can be exposed to such high temperatures and which are also cheap is not an easy task. Liquid metal (sodium, lead, etc.) receivers and graphite receivers have been proposed, among other possibilities. There is a complex trade-off here [1: 21], which encourages the search and trial of different technological alternatives [2: 26–27].

Finally, the HTF could be water/steam, which suffers from control problems since it operates in a vertical position [25: 91], molten salts, or pressurized air (which is not toxic, corrosive or inflammable, but whose heating requires a large surface) [30: 184–185].

A solar tower without energy storage shows annual capacity factors close to 25%, and it could potentially operate for ≥65% of the year with TES. Solar towers need around 2.4 m^3 of water for wet cooling per MWh generated (and ~0.15 m^3/MWh with dry cooling). Finally, lifecycle GHG emissions of solar towers show the following approached values (minimum, 25th percentile, median, 75th percentile and maximum, respectively): 11, 23, 29, 39, and 56 g CO_2 eq/kWh [7: 372 and 982].

2.3 Linear Fresnel

The Fresnel-type collector can be considered a variant of the parabolic trough [3, 18, 20: 56, 32: 2502–2503]. In this case, flat and relatively narrow mirror stripes are laid side by side one from another in the ground (although they can also be located above the ground). Glass mirrors, about ~3 mm thick, have a layer of silver as a reflecting material. With the aim of tracking the sun, mirrors can be slightly laterally turned by a mechanical bending according to the ring design (there is a circular support in each extreme of the mirrors stripes) or the bench bar design (the mirrors are disposed on rotating bars). The mirrors focus a receiver in which an absorber tube runs through. The receiver is appropriately placed above an elevated structure over the solar field. There are several technical variants of the receiver. One of the most interesting options is to locate the absorber receiver in the focus of a secondary concentrator which covers it. This leads to an increase of the solar irradiation intercept factor, reaching temperatures above 300 °C within the created cavity [11: 21–23].

The optical performance is lower than in a parabolic trough, although the thermal efficiency (45–55%) and the solar-to-electricity efficiency (~18%) are similar. What makes this technology interesting is the comparatively lower cost of the equipments of the solar field. The reason is that the flat mirrors are easy to manufacture, that the structures which support them are simpler and that the number of high-pressure joints of the receiver tube is lower. Furthermore, it leads to considerable time savings in the assembling of the solar field and the O&M activities are also simplified. Another economic advantage is that the Fresnel devices entail a dense occupation of the land, which implies a smaller sweep of the solar field per MW. However, due to their lower optical efficiency, they require a 1/3 more surface area in order to collect the solar irradiation and be able to raise the temperature of the working fluid to levels similar to parabolic trough collectors [22: 38–39]. The greatest disadvantage of a linear Fresnel plant is the optical losses. Therefore, the advantage in terms of generation costs per kWh goes down with respect to a parabolic trough plant, with increasing DNI levels [11: 31]. Regarding water consumption for wet cooling, the linear Fresnel technology requires >3.5 m^3/MWh (the highest value for CSP). Finally, lifecycle GHG emissions of linear Fresnel plants show the following values (minimum, median and maximum, respectively): 16, 18, and 22 g CO_2 eq/kWh [7: 372 and 982].

At the end of the past century, the basis of linear Fresnel technology was perfectly defined. However, its diffusion was slow due to its lower performance under high DNI levels. Some pilot plants use direct generation steam, that is, without an HTF medium, since they have been designed to obtain steam which complements coal, gas, or biomass thermal power stations [3, 11: 36–37].

2.4 Paraboloidal Dish with Stirling Engine

The fourth solar thermal electric technology is the paraboloidal dish solar concentrator endowed with a heat engine in its focal point. The set is called solar dish/Stirling engine system. In this case, solar irradiation is collected and concentrated on the point in which the engine is located. In this engine, thermal energy cyclically expands and compresses a given quantity of hydrogen, or helium, which reaches temperatures above 700 °C. This thermodynamic cycle moves a piston, and thus, thermal energy is converted into mechanical energy. Finally, pistons move a crankshaft connected to an electric generator and mechanical energy is transformed into electricity.

The best solar concentrator shape is a paraboloid of revolution. For economic and manageability reasons, complete disks of a uniform surface are not used. The alternative is to approximate this shape with multiple, spherically shaped mirrors supported with a tied-up structure. The surface for capturing the rays is thus fragmented. The supporting structure has to maintain the sun concentrator stabilized under any atmospheric condition, while moving it according to the sun's transit. The two-axis tracking goal is accomplished by an azimuth–elevation device or by a polar tracking (only for small dish/engine systems). For the above-mentioned reasons, the supporting structure is relatively complex and expensive.

Concentrator dishes have a reflective surface of aluminum or silver, deposited on glass or plastic. They are approximately 1 mm thick. As mentioned above, mirrors should have low-iron contents in order to improve reflectance, with values in the range of 90–95%.

Given that the cost of the disk (single or fragmented) and the complexity of handling it grow substantially with the size, there are limits to its size and, thus, to its capacity. A standard commercial size would be a collector from 10 to 12 m diameter which is able to generate 25 kW. However, much larger disks have been built[4] and several disks can be connected to each other, leading to solar fields of MW size.

The Stirling engine takes its name from the Scottish Reverend Robert Stirling (1790–1879), who put in practice the ideas of George Cayley (1773–1857) about a mechanical device moved by the force of enclosed hot air subject to an external source of heat. Without entering into details, in its simplest version, the Stirling engine consists of a cylinder which contains gas and a couple of pistons placed one over another, which go up and down with some delay. The use of an external source of heat (e.g., direct solar irradiation) helps expanding the gas by pushing the pistons. Its connection to an ingenious gear ensures that, while the upper one returns to its original position, it pushes the gas to the cold zone of the cylinder. Meanwhile, the lower piston starts to go up due to the reduction in pressure, pushing gas to the cavity where it will be heated again. The obtained mechanical energy can be employed for different types of uses.

Stirling started to design his engine in Edinburgh in 1816 as an alternative to the steam engine. He patented a prototype in 1827 [21: 168]. The largest model he built had a capacity of about 21 HP. However, none of the engines were commercially produced. The applications came with the Stirling engine version of John Ericsson (1803–1889), who installed it in some boats, such as the *USS Monitor* (1862) of the Union Navy. The engines designed by Ericsson, however, had a relatively small source of heat, and this explains its low power. The German Wilhelm Lehmann and later the American Alexander K. Rider presented improved versions in the last third of the nineteenth century. Rider and Ericsson founded the Rider–Ericsson Engine Co., which remained in operation until 1930. Unfortunately, the Stirling engines were greatly super-seeded by internal combustion and electric engines. Although it was used in some trucks and buses, they were relegated to being only a mechanical curiosity. The advantage of those engines is that their physical wear and tear is minimum, which implies a low maintenance. Stirling receivers are about 90% efficient in transferring the thermal energy delivered by the solar dish.

In the case of the thermo-solar application, there are two options: the kinematic Stirling engine, in which hydrogen is the working fluid subject to expansion and compression cycles, and the free piston engine, which uses helium. The first one is more efficient, whereas the second requires less maintenance. With respect to the engine receiver, the most common is the direct illumination receiver. There are also indirect receivers in which an intermediate heat transfer fluid is used. Directly illuminated receivers are capable of absorbing high levels of solar irradiation (around 75 W/cm^2). The set engine/generator is sustained in the focal point of the disk by a strong metallic arm. As of today, there are few manufacturers of Stirling engines and also few manufacturers of solar disks, which can partially explain their relatively high costs.

---

[4]Lovegrove et al. [15] mentioned a dish made up of 380 mirror panels which, once they were appropriately nested, they made up a paraboloidal dish of 25 m diameter and 500 m^2 collector surface.

A solar dish installation also has some ancillary equipment, such as controls of dish activity and tracking. However, it does not require a typical cooling system because Stirling engines transfer waste heat to the environment through a radiator.

Land requirements for solar dish systems are around 1.2–1.6 ha/MW. Water is only required for mirror washing, but not for engine cooling (<150 L/MWh). With respect to the lifecycle GHG emissions, dish/Stirling systems show the following values (minimum, 25th percentile, median, 75th percentile and maximum, respectively): 7, 11, 12, 32, and 89 g CO_2 eq/kWh [7: 372 and 982]. Furthermore, another advantage is that a solar dish/Stirling engine does not require unique and rare materials. However, the most important advantage of a solar dish and Stirling engine is its high energy efficiency: According to estimations, the conversion factor of solar irradiation to electricity is between 25 and 31.25% [22: 41] and could be even higher (see below). This is due to the high temperatures being reached: >800 °C, due to the degree of concentration of the solar rays (from 1000 to 4000 times). It is also due to the absence of fluids making long journeys. The cost of generation, however, is relatively high although (as with the rest of CSP technologies) it all depends on the number of annual full-load hours, the investment costs, and the interest rates [14: 211] (see Chap. 4).

2.5 Other Technologies

To end up, there are two other solar thermoelectric technologies: the solar updraft tower plant (or solar chimney) and the solar pond. In the former, the incident solar irradiation leads to an air convective flow inside a circular greenhouse space (with a translucent roof), covering some square kilometers. The generated heating air rises up throughout a vertical tower (of hundreds of meters high) located in the middle of such a collector, after a turbine generator located in the base of the tower is moved. In theory, such plants could operate 24/24 h if tight water-filled tubes were placed under the roof to deliver its heat at night [28: 10]. A first experimental solar chimney of 50 kW was deployed in Manzanares (Ciudad Real, Spain) in 1981 and was dismantled in 1989 after the damages caused by a storm. Years before, the Mildura project (Victoria State, Australia) consisted of a solar chimney of 1000 m high with a diameter of 130 m. It had a greenhouse space in the ground (of 38 km^2) and was able to generate 200 MW [19: 27]. Unfortunately, the project was abandoned in 2006. Some other projects have been announced, but none of them have been executed. Therefore, it is a technology which has not gone beyond the demonstration stage.

In the solar pond, dense salts are confined in its bottom, whereas water at ambient temperature forms a layer in its upper part. The water with salts cannot go up due to its high density[5], and thus, given the intense solar irradiation, the lower part of the pond reaches a temperature of 90 °C. Conduits located in the bottom of the vessel allow bringing this hot fluid to a heat exchanger, where the steam generated can

[5]The absence of wind is also required.

move a turbine. While there have been several experiences of solar ponds in Israel, Australia, and USA, their performance remains very low ($\leq$2%) [8: 56–57, 9].

2.6 Thermal Energy Storage Systems

Storing electricity is technically complex and expensive today, even at a small scale. However, storing heat is easier and cheaper. CSP plants manage large amounts of heat, which is generated during the hours of solar irradiation and is susceptible of being conveniently stored for electricity generation during the night [16, 20: 58–61]. Indeed, thermo-solar plants without storage can operate in a year at about ~25% of their nominal power, but this figure progressively rises with increasing TES capacity. The goal is to finally achieve a non-stop activity all around the year. In such a case, CSP will be an enduring dispatchable technology, which is better than other conventional thermal technologies which have significant interruptions in their activity due to breakdowns or fuel recharging.

There are several alternatives to store heat from the solar field. One option is to store the heat transfer medium, currently thermal oil or molten salt, and perhaps in the future steam or phase change materials [31: 177] in two tanks: the hot and the cold tanks. The other one is mass storage, whereby the heat is injected into stationary blocks of materials. Concrete, certain rocks, and ceramic materials are usually used for this purpose.

There are several designs regarding the use of the two tanks which make up the TES. One possibility is that, during the hours with solar irradiation, a specific circuit directs part of the oil heated in the solar field toward the hot tank. If molten salts are stored in this tank, then there is an interchanger of heat oil/molten salts. The accumulated heat is used at night; that is, the synthetic oil is heated, inverting the heat exchange process, before its energy is transferred to the steam which moves the power block. After the partial loss of heat, the oil returns to the hot tank where it is heated again. In this cycle, the partially cooled molten salts are stored in the cold tank. These salts will recover their temperature in the following day by absorbing the heat of the oil heated in the solar field. The cold tank then becomes the hot tank, and the cycle starts again. It should be noted that the use of molten salts rather than synthetic oils as HTF avoids the exchange of heat with the deposits, and given its capacity to achieve high temperatures, the deposits can be of a smaller size for the same thermal content. This advantage has favored the use of molten salts, although a certain degree of hybridization should be maintained in order to avoid their solidification. In any case, whatever the working fluid is, it has to serve the power block, and thus, a sufficient amount of HTF circulates in the solar field for both tanks.

It should be indicated that, in some cases, the bottom of the tanks contains materials with a large capacity to retain heat. In this case, the molten salt which enters and leaves the tank only occupies part of the tank. Tanks can store energy for several days with the help of appropriate insulation.

Molten salts are an efficient and inexpensive medium to store thermal energy. For years, it has been used in chemical and metalworking industries as a heat-carrier fluid. The well-known nitrate salts can be used, but not only. Lithium salts (a non-abundant material) as well as calcium nitrate salts (with a low point of solidification, but less stable and very viscose) can also be used. However, the most promising materials to store heat are the organic and inorganic substances known as phase change materials, which have a high temperature of fusion. Their name is due to the quality of changing from solid (their state at low temperatures) to liquid (after absorbing heat). Therefore, they can be stored in order to absorb or to release heat, accordingly. Paraffin is an example of an organic phase change materials, and hydrates salt is an inorganic one. They can store 5–14 times more heat per unit volume than common heat storage materials such as water. The phase change materials are known since the late 1800 s as a medium for thermal storage applications. They perform very well in small containers, or in larger containers which are divided into cells of appropriate geometry. Among the main features of these materials for heat storage in CSP plants, their high volumetric latent heat storage capacity, compatibility with conventional materials of construction, chemical stability, safety (non-toxic and non-flammable), low cost, and recyclability should be considered.

Another medium to store heat is purified graphite which, combined with molten salt, remains in tanks. It has high mass and volumetric heat capacity, and relatively low cost.

Therefore, heat storage in a CSP plant shows two types of challenges:

- The trade-off between the size of the solar field and the requirements of the power block and the TES.
- The choice of the most convenient material which will be accumulated in the heat tanks.

2.7 Solar Thermal Hydrogen and Desalination Processes

Thermo-solar generation allows for the generation of co-products together with electricity. Perhaps the two most relevant are hydrogen for industrial uses and for fuel cells vehicles and, especially, the use of plants for water desalination (for agricultural use and human consumption) [27]. There are several possibilities with respect to hydrogen [10: 839–840]:

- To use the electricity of the plant to break the water molecules.
- For solar thermal electrolysis. This requires a temperature above 700 °C.
- Cracking methane using the heat of the solar field, although this generates CO_2 emissions.
- Combining zinc with high-temperature steam to obtain zinc oxide and hydrogen.

All these processes require large improvements in terms of efficiency and cost reductions. Maybe those installations connected to CSP plants can be deployed by

2030, under the assumption that hydrogen becomes the usual fuel of transport vehicles.

Two main methods exist to obtain large quantities of drinking water from brackish water, which is normally coming from the sea: thermal (whose most common technology is multi-effect distillation) and mechanical, where reverse osmosis stands out. In general, in the first case, the temperature of the feedwater is raised until sufficient steam at the desired temperature and pressure is obtained in a cavity (the evaporator). After going through a demister (a special container in which the liquid droplets are removed from a vapor stream), this steam is directed to a condenser. The distilled water obtained is then subjected to a mineralization process. The thermal plants usually have different evaporator/condenser stages set up serially to improve the efficiency of the process. On the other hand, reverse osmosis enters the saline solution at pressure (frequently from 60 to 70 bar) in cylindrical containers which have a semipermeable membrane which lets pass pure water but does not allow saline ions and the rest of substances which it may contain to go through.

The membrane is a set of several layers of cellulose triacetate and polyamide which filter the water through tiny pores (from a few microns to hundreds of nanometers[6]). Its useful life is only between three and five years since the deposition of particulates or biological agents, pH, etc., deteriorates them. In order to mitigate this problem, the feedwater is previously microfiltered, although the oldest plants undergo chemical treatment. In both cases, together with salt water, brine (or concentrated salt solution) is obtained. Around 2/3 of the current installed desalination capacity uses the reverse osmosis technology, 30% use steam conversion, and the rest uses other technologies (see [22: 2–22]). In both cases, the energy consumed represents between 30 and 40% of the cost of the desalinated water.

It is currently assumed that a CSP plant could obtain around 15 m^3/year of desalinated water per m^2 of collector surface [5: 226]. Regarding its cost, whereas desalinating water using fossil fuels would cost more than \$1/m^3, the cost could have gone down to between \$0.25 and \$0.3/m^3 in 2020 if the possibilities offered by solar thermoelectric generation had been used [26: 137–144, 153, 170].

Desalination requires intensive maintenance (in order to control corrosion, the accumulation of diverse salts in the equipments, etc.) and a lot of energy. This energy is needed in order to transfer brackish waters into steam or to boost it at pressure through the membranes as well as to move a complex system of pumps and auxiliary equipment. Steam may come from a conventional boiler or it may come from the deviation of part of the thermal energy of a CSP plant (i.e., the thermal fluid transfers heat to the feedwater which is then converted into steam). This requires a temperature below the one reached by the HTF when circulating in the solar field. The plant can also provide electricity for the needs of the process. If the desalination process is reverse osmosis, then the CSP plant only provides electricity in order for the equipment to maintain the pressure at which the system operates. In this case, given

[6]A micron (1 μm) is equivalent to a thousandth of a millimeter and comprises 1000 nm. For example, algae usually have a size between 5 and 100 μm, bacteria are between 0.4 and 30 μm, viruses are between 0.01 and 0.1 μm, and dissolved salts (Ca, Na, Mg) are between 0.0001 and 0.001 μm.

that the lack of activity degrades the membranes, the possible intermittency of a CSP plant becomes a problem.

Obviously, solar thermoelectric plants which generate electricity together with the support for desalination would be located in coastal areas or suitable locations for the intake of brackish water. This would facilitate obtaining freshwater for the power block cooling purposes and to clean the collectors, as well as for human consumption and agricultural uses. In this case, the available freshwater would be a key resource to substantially improve the living conditions of the population located in an arid or desert region.

References

1. ATKearney (2010) Solar thermal electricity 2025. Clean electricity on demand: attractive STE cost stabilize energy production. A.T. Kearney Inc., Estela. Available at http://www.estelasolar.eu. Accessed May 2012
2. Behar O et al (2013) A review of studies on central receiver solar thermal power plants. Renew Sustain Energy Rev 23:12–39
3. Bockamp S et al (2003) Solar thernal power generation. Berlin, Fresnel-Collectors Project, E.On/Fraunhofer ISE/PSE/DLR. Available at https://pdfs.semanticscholar.org. Accessed May 2008
4. Carter NT, Campbell RJ (2009) Water issues of concentrating solar power (CSP) electricity in the U.S. Southwest. In: CRS report for congress. Available at http://www.crs.gov. Accessed May 2014
5. Compain P (2012) Solar energy for water desalination. Proc Eng 46:220–227
6. Drury E et al (2012) Solar energy technologies. In: Hand MM et al (eds) Renewable electricity futures study. Volume 2: renewable electricity generation and storage technologies, pp. 10-1–10-60. Report NREL/TP-6A20-52409. National Renewable Energy Laboratory (NREL), Golden, CO (Chapter 10). Available at http://www.nrel.gov/analysis/re_futures. Accessed July 2014
7. Edenhofer O et al (eds) (2012) Renewable energy sources and climate change mitigation. Special report of the intergovernmental panel on climate change. Cambridge University Press, Cambridge
8. Everett B (2004) Solar thermal energy. In Boyle G (ed) Renewable energy. Power for a sustainable future, 2nd edn. Oxford University Press, Oxford, pp 17–64
9. El-Sebaii AA (2011) History of the solar ponds: a review study. Renew Sustain Energy Rev 15:3319–3325
10. GEA (2012) Global energy assessment. Toward a sustainable future. Cambridge University Press and IIASA, Cambridge, Laxenburg
11. Günther M (2011) Advanced CSP teaching materials. Linear fresnel technology. EnerMENA and DLR (Chapter 6). Available at http://www.energy-science.org. Accessed Apr 2017
12. Günther M et al (2011) Advanced CSP teaching materials. Parabolic trough technology. EnerMENA and DLR (Chapter 5). Available at http://edge.rit.edu. Accessed Apr 2017
13. Hinkley J et al (2011) Concentrating solar power—drivers and opportunities for cost-competitive electricity. CSIRO. Available at http://www.csiro.au. Accessed June 2013
14. Laing D, Weinrebe G (2007) Dish/Stirling systems. In: Kaltschmitt M et al (eds) Renewable energy. Technology, economics and environment. Springer, Heidelberg, pp. 203–212
15. Lovegrove K et al (2011) A new 500 m^2 paraboloidal dish solar concentrator. Sol Energy 85:620–626
16. Madaeni S et al (2012) How thermal energy storage enhances the economic viability of concentrating solar power. Proc IEEE 100(2):335–347

17. Mai T et al (2012) Appendix C. Estimation of life cycle greenhouse gas emissions. In: Mai T et al (eds) Renewable electricity futures study. Volume 1: exploration of high-penetration renewable electricity futures. Report NREL/TP-6A20–52409. National Renewable Energy Laboratory (NREL), Golden, CO, pp C-1/C-11. Available at http://www.nrel.gov/analysis/re_futures. Accessed July 2014
18. Mills DR (2000) Compact linear Fresnel reflector solar thermal powerplants. Sol Energy 63(3):263–283
19. Mills D (2004) Advances in solar thermal electricity technology. Sol Energy 74:19–31
20. MIT (2015) The future of solar energy. An interdisciplinary MIT study. Massachusetts Institute of Technology, Cambridge. Available at http://mitei.mit.edu. Accessed May 2015
21. Montgomery SL (2010) The powers that be. Global energy for the twenty-first century and beyond. The University of Chicago Press, Chicago
22. Palenzuela P et al (2015) Concentrating solar power and desalination plants. Engineering and economics of coupling multi-effect distillation and solar plants. Springer, Berlin
23. Perales T (2011) Instalaciones termosolares. Tecnologías, Sistemas y Aplicaciones. Creaciones Copyright, Madrid
24. Pitz-Pall R et al (2005) European concentrated solar thermal road-mapping. WP 3 deliverance no 7. Ecostar Team-DLR. Available at http://www.solarpaces.org/library. Accessed Apr 2010
25. Pujol R (2009) Le retour d'un acteur français dans le secteur de la technologie du solaire à concentration. Réalités Industrielles, Nov 2009, pp 89–95
26. Trieb F et al (2005) Concentrating solar power for the mediterranean region. Final report. German Aerospace Center-Institute of Technical Thermodynamics/Federal Ministry for the Environment, Nature Conservation and Nuclear Safety, Stuttgart. Available at www.dlr.de. Accessed Apr 2010
27. Trieb F et al (2007) Concentrating solar power for seawater desalination. In: Paper presented at Middle East North Africa renewable energy conference) (MENAREC), 4th edn. Damascus, Syria 20–24 June 2007. Available at http://citeseerx.ist.psu.edu. Accessed Aug 2011
28. Viebahn P et al (2008) Final report on technical data, costs, and life cycle inventories of solar thermal power plants. Deliverable no 12.2-RS Ia, project no 502687 NEEDS/Sixth FP. DLR, CIEMAT. Available at https://www.solarthermalworld.org. Accessed Nov 2012
29. Wienrebe G (2007a) Solar tower power stations. In: Kaltschmitt M et al (eds) Renewable energy. Technology, economics and environment. Springer, Heidelberg, pp 181–193
30. Wienrebe G (2007b) Parabolic trough power plants. In: Kaltschmitt M et al (eds) Renewable energy. Technology, economics and environment. Springer, Heidelberg, pp 194–203
31. Wienrebe G, Ortmanns W (2007) Principles. In: Kaltschmitt M et al (eds) Renewable energy. Technology, economics and environment. Springer, Heidelberg, pp 171–181
32. Xie WT et al (2011) Concentrated solar energy applications using Fresnel lenses: a review. Renew Sustain Energy Rev 15:2588–2606

Chapter 3
Short History, Recent Facts, and the Prospects of Concentrating Solar Power Generation

This chapter deals with three important issues related to the history of CSP development, namely the early steps and pioneers of thermo-solar technology (Sect. 3.1), the CSP diffusion facts from 1980s to today (Sect. 3.2), and the drivers and barriers to its deployment (Sect. 3.3).

3.1 Early Steps

From ancient times, humankind has recognized and taken advantage of the role that solar energy in the form of light and energy has for its survival. On the one hand, for thousands of years, the symbols, myths, and the pantheon of deities have been full of reminiscences to the presence of the sun and solar activity. On the other hand, human communities have used solar irradiation for pragmatic purposes: heating dwelling spaces (private houses and public baths), drying materials, and objects, raising the temperature of liquids, separating the salt from the water, etc. Thus, for example, texts from the Greek and Roman ages already suggested to orient the main facade to the North, whereas the rooms had to be oriented to the South, with the aim to use the solar heat and to save wood and charcoal. Whereas, in summertime, curtains had to be deployed to restrain the entry of hot air, in the cold months the retention of solar heat was improved by closing the windows of the south façade with light-colored materials, such as mica, alabaster, or glass [13: 3–27]. In other cultures, such as the Pueblo Indians (twelfth century), towns were built to take advantage of the sun in the winter, facilitating heat retention through adobe walls and using straw in the roofs as insulation material.

The building of greenhouses was a common practice in the classical antiquity and lasted for centuries, although it was not until the nineteenth century when it reappeared strongly. For the zones of temperate climate, the British architect Humphrey Repton suggested to put a space as a greenhouse in the back of buildings, connecting

P. Mir-Artigues et al., *The Economics and Policy of Concentrating Solar Power Generation*, Green Energy and Technology,
https://doi.org/10.1007/978-3-030-11938-6_3

it to the living room. This possibility was indebted to the advances in the manufacturing of larger windows panes since the end of the seventeenth century [13: 47–53]. Also in the 1930s and the later postwar period, different architects proposed buildings where the thermal use of solar irradiation would be a priority. Notwithstanding, all those designs were quickly forgotten when energy became cheaper.

Since immemorial time, the use of convex lens made up of crystal rocks to cauterize wounds is known. For example, the Chinese and the Greeks used concave reflectors to concentrate the solar rays and light a fire. Some proposed building large mirrors to use them in the battlefield (with success in the case of Archimedes against the Roman boats which sieged Syracuse) although, in reality, the methods or equipment to manufacture them at a very large size were not available [85: 267–268]. Notwithstanding, around 1515, Leonardo da Vinci designed a machine to shape and polish them [102: 36–37].

Regarding the use of direct solar irradiation to obtain hot water, a first experience worth mentioning is the one of the Swiss naturalists Horace-Bénédict de Saussure (1740–1799). In 1767, he built a box with the inner part painted in black where he fitted, one inside another, up to five square boxes of glass of decreasing size. Temperatures up to 109 °C were reached inside the box. Although he was not able to find an answer to such phenomenon, he had created the first solar oven. Today we know that the inner dark face absorbed irradiation and, although the glasses allowed the light to go through, they did not allow the heat to escape. This design was continuously improved, for example by Sir John Herschel and Samuel Pierpont Langley in the nineteenth century [13: 54–59]. Also in this century, the French physicist Claude Pouillet estimated the value of the solar constant.[1]

Solar water heaters were another application of solar irradiation. When it comes to obtaining water at low temperature for domestic uses, the activity of the inventor and manufacturer Clarence M. Kemp, from Baltimore, stands out. He patented a modern solar water heater to be placed outside of houses, which he commercialized under the name "Climax". He intended to sell it to businessmen who were home alone because their families and the domestic service had temporarily moved to the second residence. In this situation, they might have difficulties to switch on and maintain a boiler and, thus, the solar heater would be quite useful. The product was highly successful in California, where about 1600 units were sold by 1900 [13: 117–122, 50: 3–4]. Despite some improvements, the device had a problem: it was exposed to the cold during the night. In the early twentieth century, it was proposed to install the deposit inside the building and use, as a water heating system, a narrow pipe cloistered in a box between a black-painted back metal sheet and a glass covering. This new configuration, together with several other improvements, was designed by the engineer William J. Bailey who, in the first decade of the twentieth century, sold more than 4000 "Day and Night Solar Hot Water Heaters" [13: 129–141, 50: 4]. However, in the 1920s, the Californian market for solar water heaters collapsed due

[1]The instant electromagnetic radiation from the external atmosphere is 1361 W/m^2 that, once impacting the earth surface reaches an average of 1050 W/m^2 of direct sunlight and 1120 W/m^2 if diffused light is added. This latter value is the solar constant.

to the start of cheap gas extraction. Bailey then started to manufacture gas heaters. However, the sales of water heaters continued to be important in Florida until the fall in electricity prices after World War II [85: 268–270]. Since then, the setting up of huge gas distribution networks as well as the low costs of electricity has slowed down its market in high solar regions that would justify the use of water heaters.

The perception that the sun can greatly increase the water temperature, so that the obtained steam can have many uses, goes back to the designs of the Neapolitan Giambattista Della Porta (1535–1615). The device that he proposed in 1601 would be revised years later by Salomon de Caux (1576–1626). After a century and a half, the effects of concentrated solar irradiation would allow Lavoisier and Priestly to advance their investigations on gases [5]. However, Augustin Mouchot (1825–1911), a professor of mathematics at the Lycée of Tours, was the one who took advantage of the water heated beyond 400 °C by concentrated solar irradiation. Worried about the exhaustion of coal reserves, Mouchot published in 1869 the book *La Chaleur Solaire et ses Applications Industrielles*, in which he proposed solar irradiation as an alternative energy source. This was based on his experiences in previous years: He started to improve hot boxes, then he built a solar oven and a still for distilling wine into brandy and he ended up designing the (probably) first steam-driven engine fed by solar energy [13: 65–68]. Throughout this process, Mouchot realized that, in order to produce steam at the optimal quantity and temperature, the size of the collector needed to be huge, which was not operative and was too expensive. Therefore, he proposed to concentrate the solar rays with concave disks [84: 4]. In 1874, with a small support from the French administration, he presented his new solar machine in Tours. His concave collector, made up of copper plates coated with polished silver, pivoted in a way which could be oriented to the sun, although this task had to be performed manually. Furthermore, he reached the conclusion that the solar engines would only be viable in the sunny tropics. He had the opportunity to check this in Algiers (Algeria, then French colony), where he arrived in March 1877, with the support of the government. There, he tested solar ovens and stills, although his experiences with solar water pumps for irrigation were what merited more attention from the French authorities [13: 71–72]. With additional resources from the government, he built a large paraboloidal collector whose associated engine was able to pump 1900 L/h, as well as an ice-maker device. He presented all his inventions in the World's Fair of Paris in 1878. In 1882, this solar machine was converted into the first solar power printing press with the help of his assistant, the engineer Abel Pifre (1852–1928). During the festival of *L'Union Française de la Jeunesse*, the printing press, which was moved by a steam engine powered by a solar dish, would print about 500 copies per hour of *Le Journal du Soleil* [92: 5]. Two years before, however, and after two decades intensively dedicated to the use of solar energy, Mouchot, he had abandoned all the projects. He had concluded that the parabolic trough was a better surface to capture the solar rays than the collector dish. Although it was a transitory curiosity, neither private investors nor public authorities were interested in those sun machines. The latter considered that those devices were not useful beyond very sunny arid zones where, in addition, it would be very difficult to access other energy sources.[2] Mouchot and Pifre did not protect their achievements: The first solar patent of a multiple solar collectors would be awarded, eight years later, to the Italian Alessandro Battaglia (1842–n.a.). This engineer had published, in 1884, his paper *Sul modo e sulla convenienza di utilizzare il calore solare per le macchine a vapore* in which he proposed to heat a boiler with the solar rays reflected by hundreds of small flat mirrors [99: 2–3].

Another name to be remembered is the English engineer William Grylls Adams (1836–1915). He started his career in the British patent office in London, where he accumulated abundant knowledge about energy generation systems. After working as an engineer for the colonial administration in India, he published the book *Solar Heat: A Substitute Fuel in Tropical Countries* in 1878. Starting from the designs of Mouchot, he proposed to complement the burning of expensive coal with solar energy in order to generate steam [50: 5–6]. More specifically, its project consisted of flat silvered mirrors placed in a semicircle to reflect the solar rays to a stationary boiler. The mirrors would be moved manually. Despite his insistence that this installation would be able to produce steam at very high temperatures—as suggested by the small device able to power a 2.5 HP steam-driven engine which he installed at his home [92: 8]—, the colonial authorities were unfortunately not convinced. Disenchanted, he tried to manufacture and sell solar kitchens in Mumbai. Having returned to London, he became professor at King's College.

This historical review also stops with the contribution of the famous North American engineer of Swedish origin John Ericsson (1803–1889), who was an expert in machines and thermal engines. His concern about the exhaustion of coal made him being interested about the possibilities of solar energy. Although he was dedicated to several engineering fields, his *Solar Investigations* (1875) stand out. He claimed that, in very sunny regions, steam could be obtained through the concentration of solar irradiation. Then, through the use of an adequate engine, water could be pumped. He estimated that the equivalent to one horsepower could be obtained by a hundred square feet of mirrors. Ericsson built in 1884 the first known parabolic trough collector [13: 80] and designed different concave dishes and auxiliary devices, although he ended up admitting that those receivers were expensive to manufacture and complex to operate. The advantage of the gratuity of the fuel was nothing compared to the cumbersome process of building and managing the systems to capture and concentrate the solar light. Ericsson's desire to keep his designs in secret explains that the first patent of a trough was awarded to the Germans Wilhelm Meier and Adolf Remshardt in 1907 [9: 95–96, 35: 9, 84: 5–6].

The works of Ericsson inspired Aubrey Eneas, an entrepreneur born in the UK but based in Boston. In 1892, he funded the Solar Motor Company of Boston but, being aware of the need for steam-powered irrigation in the deserts of the American southwest, he relocated in Los Angeles in 1903. Since 1899, however, he started to live in Denver. In that year, he obtained what it is considered to be the first patent of a solar dish. In 1901, he moved to Pasadena (California) and contacted the owner of

[2]Only the French Foreign Legion built some solar ovens for its troops in isolated places in the desert and some Algerian settlers used solar stills in order to distill brine water [13: 75].

the first ostrich farm in the country, the Cawston Ranch. This was something exotic which attracted many visitors, although his true real objective was to supply feathers for dresses and hats. Therefore, this was a great place to show his solar motor which, endowed with a large solar dish, was able to pump more than 5 thousand L of water to irrigate about 125 ha of citrus trees. A novelty of the device was that the solar dish could automatically follow the sun's transit, although it had to be calibrated every day. The technical and advertising success led Eneas to promote the sale of the irrigation system for $2160. The first purchaser was a landowner from Arizona. Unfortunately, violent winds destroyed the collector. Some others also bought it, although in two years it became clear that the use of the solar dish was cumbersome and its design was not accurate to resist strong storms. Notwithstanding, the main handicap was its cost per HP, which was between twice to five times the cost of a conventional steam plant. He quickly abandoned the business [13: 88–89, 50: 4–5, 92: 10–11].

Another attempt of using solar irradiation to pump water in dry and remote areas was performed by Henry E. Willsie and his colleague John Boyle, in the first decades of the twentieth century. His approach was inspired by the work of Charles Tellier, a French engineer of the second half of the nineteenth century who had used low boiling point liquids for refrigeration although, in 1885, he installed flat collectors for heating domestic water in the roof of his home [92: 9–10]. Therefore, more than redesigning or increasing the size of collectors, his goal was to use chemical compounds, especially ammonia and sulfur dioxide, in order to create a cycle liquid/steam/liquid through low-temperature solar devices, more specifically hot boxes placed in the roofs of houses or stores. First in Illinois and, since 1903, in Hardyville (Arizona) both pioneers tested their equipment with success, creating the Willsie Sun Power Company. The next year a first plant was built in St. Louis, which could operate 24/24 h because it was hybrid. Shortly after, Willsie and Boyle settled in Needles (Mojave Desert, California). In a land of their own, they installed a 15 HP plant which they were improving depending on the availability of financial resources. The fourth and last version of this plant was pioneer in having an insulated tank where hot water was stored and later used to vaporize sulfur dioxide during the nighttime. This was a technical achievement which made Willsie and Boyle the pioneers of thermal storage. The correct functioning of the plant conflicted with the superior efficiency of the engine moved by artificial gas coming from burning coal [13: 91–99]. In the last patents which Willsie obtained, he added electricity generation to the uses previously suggested for the solar engines [92: 12].

The North American engineer Frank Shuman (1862–1918) deserves special attention for operating, in 1913, the first solar field of parabolic trough. The solar career of Schuman had started five years before, after winning a lot of money by inventing a shatterproof glass and having participated in other industrial projects [92: 15]. He studied the work of his predecessors in depth, which made him experiment with hot boxes and low-temperature solar devices in Tacony (Philadelphia), the place where he lived, given that the solar reflectors were not practical. Shuman was a visionary entrepreneur who, in parallel to this first solar plant, searched for investors, and created the Sun Power Company. The resources he obtained allowed him to improve his plant by modifying the hot boxes until a U-shape directed to the sun was achieved; i.e., the capturing surface was close to a parabolic trough shape. Connected to a north—south simple axis tracking system, those collectors had to be periodically reoriented [13: 102–106]. Unfortunately, such demonstration plants and the obtained patents did not allow him to convince a sufficient number of investors. He then moved to London, believing that he would attract the attention of the authorities, whose vast empire was extended over tropical regions, as well as private resources. In 1911, the Sun Power Company was split in the Shuman Engine Syndication Ltd, and the Sun Power Company, Eastern Hemisphere. In addition, it moved to Al Meadi, a town in the west bank of the Nile River, 24 km South of Cairo (Egypt). In 1912, he deployed a plant, named Solar Engine One, with a solar field made up by perfect parabolic trough collectors, following the previous advice of British engineers.[3] Those trough-shaped mirrors were made up of small panes of silvered glass supported by brass sheets. In total, five collectors supported by steel frames, and with a length of 62 m and a width of 4 m, were installed. They were oriented north/south and had a tracker mechanism. A glass-covered tube was placed along the trough focus through which water circulated. Its temperature could reach 93.5 °C. The system could generate enough steam to move an irrigation pump (of 55 HP and imported from the USA), which was able to pump 22,750 L of water per minute from the river to the adjacent fields [3: 999, 5: 14, 50: 8, 83: 21–22, 92: 16–18]. The cost per liter of water was lower than the costs of pumping burning coal, an expensive fuel because, in Egypt, it came from far away. Then, the colonial authorities offered him the possibility to build a plant in British Sudan to irrigate more than 12 thousand ha dedicated to the then highly lucrative cultivation of cotton. Shortly after, the German authorities, who had learned from the success of Schuman, called him to the Reichtag, promising him a large sum to build a plant in their domains in East Africa.[4] However, with the start of World War I, all those projects were abandoned with the hope to restart them in the future, although following the mandate of the authorities, the plant of El-Maadi was virtually dismantled [92: 25–26]. Unfortunately, Shuman died in 1918, before the war had ended. After that, the much cheaper oil displaced coal imports, which put an end to the expectations which all those African experiences had raised.

For years, nothing relevant happened in the field of solar thermoelectric generation. The exception was the first mirror dish built by R. H. Goddard, an engineer known for his works with rockets, at the end of the 1920s. It would not be until the 1950s that the new design would be gradually presented in scientific congresses and new prototypes would be built. From the early 1960s and for two decades, the initiatives of the Italian mathematician Giovanni Francia should be highlighted (1911–1980) [2, 3: 1009, 72: 24, 99: 4–5]. To start with, this professor at the University of Genoa focused his work on the linear Fresnel collector following the

[3]Among these, the physician Charles Vernon Boys, who had vast knowledge on tracking systems and parabolic reflectors, stands out [92: 21].

[4]The attention that he was getting led Shuman to speculate that it would only be necessary to cover 5 thousand km^2 of the Sahara with trough reflectors in order to generate 270 million HP, a similar power to the one provided by all the hydrocarbons extracted in 1909 [84: 6].

steps of Battaglia (mentioned above) and including the recent designs of the Soviet engineer Valentin A. Baum. After he obtained patents on solar devices in 1961 and 1962, he built the linear collector of Lacédémone-Marseilles (France) in 1963, with support from France's National Research Council (CNRS), NATO, and COMPLES (Coopération Méditerranée pour l'Energie Solaire). The improvements introduced in the aluminum reflectors and linear boiler allowed to obtain 38 kg/h of steam at 100 bar and 450 °C. This was an important achievement, although it was deemed insufficient by Francia. At least 150 bar and >500 °C would have to be achieved in order to generate electricity at a reasonable cost. This is why he deviated his attention to the point focus concentrators, known as solar towers today. Thus, he designed and built the first solar tower in Sant'Ilario, near Genoa, in 1965, followed by three additional prototypes.

Giovanni Francia also participated in other solar thermal projects. The most relevant was the solar tower Eurelios, which was located in Adrano (Sicily, Italy). This project started to be built in 1976 and was connected to the grid in April 1981 after its finalization, tuning, and test. It was built by an Italian–German–French industrial consortium and the Italian utility ENEL. It was a demonstration plant with 182 heliostats and a tower of 55 m high, which was able to generate 1 MW. The project had the support of the European Commission. Its operation left it clear that the heliostats had to be redesigned and that the plant conversion efficiency had to be increased in order to reduce the high generation costs [3: 1012–2013].

The two oil crises in the seventies radically changed the expectations of solar thermal generation and renewables in general. In the case of CSP generation, despite the importance of the previous experiences, the first great milestone was the successive parabolic trough plants Solar Electric Generating Systems (SEGS) which were deployed, since 1979, by the company LUZ International Inc. in California.[5]

Arnold Goldman (1943–2017), an engineer born in Rhode Island and graduated in the University of Southern California in 1967, started his career in the field of computer programming. In 1977, after winning a lot of money, he went to Israel. Goldman was a visionary who dreamt of building an utopian community which, based on the bible texts, he called LUZ. A particular element of this society was that it obtained his energy from the sun. Therefore, together with Patrick François, he created the company LUZ International, whose first project was a small solar energy steam generator for a kibbutz [50: 31–32]. However, he came back to the USA due to the bad perspectives of the business in Israel. He reoriented his business, being aware that the desert zones were the best to install solar plants and informed about the advantages which the administration offered to generate renewable electricity. Between 1984 and 1990, his firm raised $1 billion to shape what it is still today considered a legendary project, the SEGS.

[5]It should be mentioned that, between 1977 and 1982, parabolic trough collector prototypes had been built by the company Acurex in the USA in order to generate steam for industrial uses. Also a primer line-focusing solar power plant of 150 kW had been proven in 1979 in Coolidge (Arizona). These and other initiatives are described in Günther et al. [35: 9].

The SEGS were a set of nine plants, whose total capacity reached 354 MW.[6] The federal and state incentives provided by Energy Tax Act of 1978 and the California Public Utilities Commission of 1983 (in accordance with the Public Utilities Regulatory Policy Act, PURPA), respectively, allowed undertaking the project and its continued expansion between 1984 and 1990. Plants sold the electricity generated to the Southern California Edison utility for a 30 years long-term PPA. In the following paragraphs, the main features of each SEGS plant are discussed (more information is provided in [3: 999–1002, 35: 10–11, 81: 44–50, 92: 29–34]):

- The pioneer plant SEGS I started its activity on December 12, 1984, in Barstow, Mojave desert.[7] Its nominal capacity was 13.8 MW and occupied 90 thousand m^2. The investment cost was $62 million, that is, around $4500/kW. After going through the solar field, steam reached 297 °C and was reheated with gas until 416 °C before entering the turbine.[8]
- The building of SEGS II, with an investment of $3187/kW, a nominal capacity of 30 MW and a land occupation of 165,376 m^2 started the following year. In this case, mineral oil was substituted by synthetic oil, which allowed it to reach 316 °C when leaving the field. Its high costs discouraged the installation of a storage system, although the plant maintained the hybridization. The turbine was made up of two bodies. Neither SEGS I nor SEGS II reached an optimal temperature of the HTF when leaving the solar field, although they represented essential steps to achieve such objective.
- Plants SEGS III–V were very similar. Their main technical novelty was that they had an improved collector, which increased the temperature of the working fluid up to 350 °C when leaving the solar field. An auxiliary oil heater was added.
- With SEGS VI and VII, the improvements in the solar field continued (HTF at 395 °C) and the components of the power block were redesigned, which increased efficiency up to 23% (hybridization included). They could operate in solar mode, only with gas or with both solar and gas at the same time. It should be mentioned that each of the SEGS III to VII plants, inaugurated between 1986 and 1998, had a capacity of 30 MW. They were deployed in Kramer Junction.
- The last two plants (SEGS VIII and IX, years 1989/1990) had 80 MW each. Therefore, they were much larger than the previous ones, and their electricity generation costs were 25% lower. It should be mentioned, however, that the project developers pressed PURPA to remove the regulatory limit for plants (only plants with a capacity up to 30 MW were allowed). An improved hybridization increased

[6]This capacity would not be exceeded until 2014 by the 392 MW of the Ivanpah Solar Electric Generating System solar tower, located in Clark Mountain, also in the Mojave Desert. When writing these lines (end of 2018), the Ivanpah plant continues to be the largest in the world, although different operative problems have hindered the full use of its capacity. Notwithstanding, it does not seem that this leadership will be maintained for a long time.

[7]And, despite the fire in the TES in February 1999, which interrupted its activity temporarily, it fed electricity into the grid. All SEGS plants, with some improvements being incorporated during the years, are still in operation.

[8]All the SEGS plants had a hybridization system and could burn gas up to 25% of their electricity production, according to California regulation.

the efficiency of plants to 34.2%. They were located in Harper Dry Lake [72: 21]. Their respective solar fields covered about 500,000 m^2 each.

- There was a project for a tenth plant, which never materialized due to the loss of the economic incentives, particularly the fiscal exemptions which had been established until then by the Californian property tax.

On average, the SEGS had a capacity factor of 21.55% and entailed an investment of $ 6233/kW ($2005). As it was mentioned, the plants were hybrid, since they burnt natural gas during the night hours (this represented up to 25% of the annual thermal energy input). The working fluid used in all of them was synthetic oil, and the thermodynamic cycle was Rankine. In the 1980s, virtually all the electricity of solar origin generated in the world came from the SEGS plants, since PV generation was then in its initial stages.

The SEGS plants were not as competitive as it was expected. As mentioned above, the electricity generated was compulsorily purchased by the utilities according to a PPA, with a price per kWh which was finally paid by consumers and equalled the avoided cost of covering the peaks in demand with conventional generation. However, in 1985, the prices of hydrocarbons fell sharply and the expectation that they would be high again disappeared. The project entered into a crisis, since the plants were only competitive during the hours of high electricity demand, i.e., when there were high prices per kWh due to the massive use of air conditioning devices (which led to the activation of conventional plants, whose generation costs were comparatively higher). Unfortunately, the reduction in the price of gas in 1992, as well as the end of some fiscal incentives (removal of the 30% investment tax credit), aggravated the financial problems. Although their hybrid nature prevented the plants from being closed, expansions were postponed (up to four additional plants had been projected) [62: 1, 84: 9]. No more plants of this type were built anywhere in the world for years. In fact, during two decades, SEGS survived as the single commercial representative of the parabolic trough plant and, by extension, of CSP plants in general. This did not prevent the publication of the first strategic plant for the development of CSP generation in 1997 [116]. The first roadmap of the parabolic trough collector technology, which was by then considered to be the most promising, was drafted on the following year [89]. These reports, however, were not enough, although the publication of a much broader one in October 2003 was highly influential in the recovery of thermo-solar activity [100].

The fall in crude oil prices in 1985 slowed down research and innovation activities. Public RD&D funds for CSP, which had been quite high in the case of the USA and, to a lesser extent, in Japan between 1976 and 1985, fell. In the USA, they disappeared almost completely by the start of this century,[9] which meant that the country abandoned a technology in which it had been a pioneer. Fortunately, the pressures to definitively cancel any support for thermo-solar RD&D were disregarded by the Department of Energy. Thus, in 2005, the DOE created the "1603 Loan Guarantee Program" to encourage investments in renewable plants, which led to

[9]In fact, there wasn't any provision of RD&D for CSP technologies by the Department of Energy (DOE) in 2004.

the inauguration of the Saguaro demonstration parabolic trough plant of 1 MW of capacity in Tucson, Arizona, the next year. In 2007, after 17 years of paralysis, the parabolic trough plant Nevada Solar I entered into operation. The funds for RD&D returned in these years, although their relevance was not as high as 30 years before. The level of expenditure never recovered in Japan. Other countries are worth mentioning in this context (albeit with modest figures), including Italy (especially during the first decade of this century), Spain, Germany, and Switzerland (see Chap. 6 for further details). Patent applications also fell sharply in the mid-1980s, although, in general, they would recover by the start of the century [10: 2448–2450]. However, the experience with SEGS plants would have a huge impact on the later developments of the sector. Thus, as reported by De la Tour et al. [19: 19–20],

- When the Israeli Solel Solar Systems bought the LUZ assets in 1992, the knowledge in making collectors was also bought. For years, this firm and Schott were the only two manufacturers of collectors.[10]
- Acciona Solar Power (created in 1997) hired some key executives from LUZ.
- Solarmundo, specialized in Linear Fresnel technology, was co-funded by the former president of LUZ.

There were many other channels of the dissemination of LUZ's know-how.

After the first Italian prototypes of solar tower, the proliferation of this type of projects took place in the first half of the 1980s, although they would continue to be experimental ones. The following are worth mentioning: the solar tower CESA-1 of 1.2 MW, projected since 1979 and inaugurated in 1983 in Tabernas (Almería, Spain)[11]; the 1000-kW solar furnace, called Héliodyssée Grand Four Solaire, built in Odelló de Cerdanya (Odeillo) in 1975 and the Themis tower demonstration plant of 2.5 MW in Targasona (Targassonne)[12] (both in French Catalonia) in 1982; and the plants SPP-5 of 5 MW in Crimea (former USSR, 1986) and Sunshine of 1 MW in Nio (Japan, 1981). In total, they added up to 23 MW, most of which would not be active before 1990. However, the most important projects would be the towers Solar One, of 11.7 MW, and the Solar Two, of 10 MW, both in Barstow (California). Solar One was built in 1981, with 1818 heliostats of 39.3 m^2 each and oil and water as HTF, under the initiative of the Department of Energy, several utilities and private firms and the presence of research centers. It also had a TES tank filled with rock and sand. Although this demonstration plant would produce electricity below the initial expectations (mostly because of the available DNI) and would have considerable problems before its definitive closure in 1988 (which left it clear that steam should be replaced by molten salts), the accumulated experience contributed to improve successive plants. It would revive in 1996 as Solar Two, with a capacity of 10 MW. This required more mirrors, a molten salt receiver, and other improvements

[10]Solel was bought by Siemens in October 2009, but it was definitely closed three years later because Siemens, eager to exit the solar business, didn't find any buyer.

[11]An important step forward was the building of the experimental solar thermoelectric complex Plataforma Solar of Almería, which is still active since 1979.

[12]This solar tower had 201 heliostats and a tower of 100 m. It was operating from 1983 to 1986, when it was dismantled due to its high generation cost [90: 93].

[3: 1013]. The investment exceeded $30 million and was covered by a public–private consortium. Its complete construction would be delayed two years and would be operative as a power plant from February 1998 to April 1999. This short lapse showed the technical feasibility of this design and its economic possibilities (conditioned by an increase in size).

The first commercial solar tower (PS 10) was inaugurated in 2007 in Sanlúcar la Mayor (Spain). Notwithstanding, the pilot plant Solar Three, or Gemasolar, of almost 20 MW, 2650 heliostats and a solar field with a surface of 195 had been erected some time before. This plant allowed testing the improved designs with respect to Solar One and Two.

The Linear Fresnel collector would undergo a direr fate. It took its name from the lents for lighthouses developed by the French physicist Augustin-Jean Fresnel (1788–1827). After the aforementioned pioneer plant erected by Giovanni Francia, there would not be another experimental one until 1991 in Israel. Two years later, the University of Sydney designed and then patented a variant, called Compact Linear Fresnel Collector, that is, a field with mirrors of alternated inclination which were able to focus two insulated steam tubes at the same time. It would not be until 1999 that the Belgian Company Solarmundo would build a prototype plant of a respectable size, since it had a collector width of 24 m and a reflector area of 2500 m^2. In 2003, another experimental plant, of about 1 MW, would be built by the Australian company Solar Heat and Power (then Ausra, which was in turn bought by the big French firm Areva in 2010) to feed steam directly into the 2 GW coal-fired power station located in Liddell (Hunter Valley, New South Wales) [16: 30, 18: 15–16, 34: 8]. Another plant was projected near the coal plant Stanwell Power Station in Rockhampton (Queensland, Australia) [11]. However, it all seems to indicate that this project would be postponed.[13] Meanwhile, between 2005 and 2008, Ausra expanded the plant in Liddell up to 9 MW. In these same years, the pilot plant Fresdemo, of the Plataforma Solar de Almería (Spain), would be in operation. In this 800 kW installation, steam was generated at 450 °C. In March 2008, Ausra started the building of Kimberlina Solar Thermal Energy Plant, in Bakersfield (California). This Fresnel plant, which was connected to the grid in October 2008, had a capacity of 5 MW. The first European commercial plant would start operation in Murcia (Spain) in the next year. The plant, which was built by Novatec Biosol, had a capacity of 1.4 MW. By the end of the past decade, all the linear Fresnel plants in the world had an aggregate capacity of 6.4 MW. There are currently 192 MW of linear Fresnel plants in operation or in the final stage of construction. However, the capacity of ongoing projects is almost twice such figure (more information is provided below).

Regarding the solar dish with a Stirling motor, the first demonstration plant of 25 kW of capacity was erected in Southern California during the first half of the in the early eighties by the companies McDonnell Douglas Aerospace Corporation,

---

[13] Although the authors were not able to directly verify this, only five thermoelectric plants appeared to be in Australia at the end of 2017, as included in the database https://www.nrel.gov/csp/solarpaces/by_country.cfm. Four of these were solar tower plants and the aforementioned Liddell Linear Fresnel Power Station, which was enlarged to 3 MW.

Advanco Corporation, United Stirling AB, and the public institutions NASA's JPL and the DOE. The plant showed numerous technical problems. In addition, the first of the aforementioned companies would deploy more dish/Stirling prototypes until the middle of the decade, when its solar activity would cease. Simultaneously, German project developers deployed, in Saudi Arabia, two solar dishes of 17 m of diameter and a total capacity of 50 kW [3: 1017].

The American company Cummins Engine Company tried to commercialize this technology in the 1990s, with the support of Sandia National Laboratories and NREL, although, in 1996, Cumming abandoned due to the difficulties in solving the technical problems of its free-piston Stirling engine. However, three years before, the Science Applications International Corporation had demonstrated the technical feasibility of a 20 kW solar dish in Golden (Colorado). This firm would deploy more experimental plants until the end of the decade [3: 1018]. There were other projects at the beginning of the century, including the North American Boeing/Stirling Energy Systems Dish Engine Critical Components and the SAIC/TM, as well as the European DISTAL I and II, located in the Plataforma Solar de Almería [72: 23]. Finally, the Maricopa Solar Power plant, with a capacity of 1.5 MW, located in Peoria (Arizona), and deployed in 2008, should be mentioned. This project relied upon the experience which had been accumulated by the Sandia laboratories (USA) some years before. In September 2009, the plant started its commercial activity. It had 60 dishes with a 26% efficiency. Unfortunately, its life was very short: from the beginning of 2010 until September 2011 when its project developer, the company Stirling Energy Systems, filed for bankruptcy.

All the aforementioned plants, despite being closed a few months later, proved to be key experiences in assessing the performance of the technology, while simultaneously contributing to the training of technicians. And they left it clear from the start that a solar dish with a Stirling engine could reach high solar-to-electricity efficiencies. A first milestone was the 31.25% efficiency that a prototype with six dishes and 150 kW of capacity reached in the Sandia's National Solar Thermal Test Facility (New Mexico, USA) in the middle of the past decade. To our best knowledge, the record is held by the prototype located in the Kalahari Desert (Northern Cape Province, South Africa), whose designers stated that they had achieved a 34% conversion rate [4].

Research on solar dishes has never stopped. An interesting variant has been the solar dish made up of many mirrors with a total aperture area of almost 500 m^2, which concentrates the solar rays (up to 2240 times, which entails capturing >90% of solar energy) in a cavity where the feed water is transformed into steam (up to 535 °C) [12]. This is a direct steam generation plant which moves a four-cylinder steam engine, rather than a Stirling engine, with a nominal capacity of 50 kW. However, the large dimension of the disk (with a diameter of 25 m) and its huge weight (19.1 metric tons) make its operation a challenging task.

The solar chimney was another technology under experimentation in the 1980s. A prototype of 50 kW of capacity was located in Manzanares (Ciudad Real, Spain), which lasted until 1989 (see above). Many solar ponds were also built in Israel, Australia, and the UAE, although virtually all of them closed by the end of that decade.

3.2 A Slow Growth

Leaving aside the more remote pioneers, whose trials would lead to the Shuman plant, the recent history of the thermo-solar sector covers an interval of about 40 years. In this period, which started in the second half of the seventies, the sector has grown in an uneven manner, that is, with a lot of activity in some short periods (and few places) and with a stagnation (together with some phases of a lower activity) for some years. This singular trend reflects the vicissitudes of the energy market and the regulatory ups and downs. The data underlying this evolution are presented in this section, which emphasizes the revitalization of the sector since 2007.

According to the data presented in Fig. 3.1, where the columns display the annual installed capacity and the rows show the accumulated capacity, the combination of parabolic trough, solar tower, and other CSP projects displays a historical evolution with three stages (adapted from data published at [17, 27, 62, 78]):

- The thermoelectric solar generation capacity started to expand in 1984, due to the successive expansions of SEGS in the Californian desert. This was a set of nine plants which would reach about 350 MW in 1991. This capacity would remain constant for 15 years. In reality, the SEGS would be the only commercial CSP initiative for more than two decades.
- Since 2007, the installed capacity gradually increased up to more than 1 GW in 2010 [10: 2443]. This stage ended in 2013 with the stoppage of the deployment of all the new capacity in the country which had led it (Spain). At this stage, the serious crisis of the European solar sector in 2012 was felt. Together with the disappearance of numerous PV companies, Solar Millennium, a leader manufacturer of components for solar thermoelectric plants, went bankrupt.
- In 2013, the leadership would return to the USA. The world accumulated capacity, which almost reached 3.5 GW in that year, increased gradually, except in 2015 and 2016. By the end of 2016, the installed capacity was between 4889 and 5193 MW, depending on the source.[14]
- For the immediate future, a clear upturn of the installed capacity, with new main actors (Morocco, South Africa, China, Saudi Arabia, etc., see below) is expected.

Leaving aside the expectations for the future, given that the accumulated installed capacity in the end of 2017 was around 6126 MW, the compound annual growth rate of installed capacity since 1984 has been 20.3%.

[14]The sources always show discrepancies due to the ambiguity in the declaration of intent, the completion of the preliminary drafting of a project, the choice, and authorization after the corresponding administrative process, the moment when construction starts, the testing phase and the definitive connection to the grid. This is a long period which is subject to all types of delays.

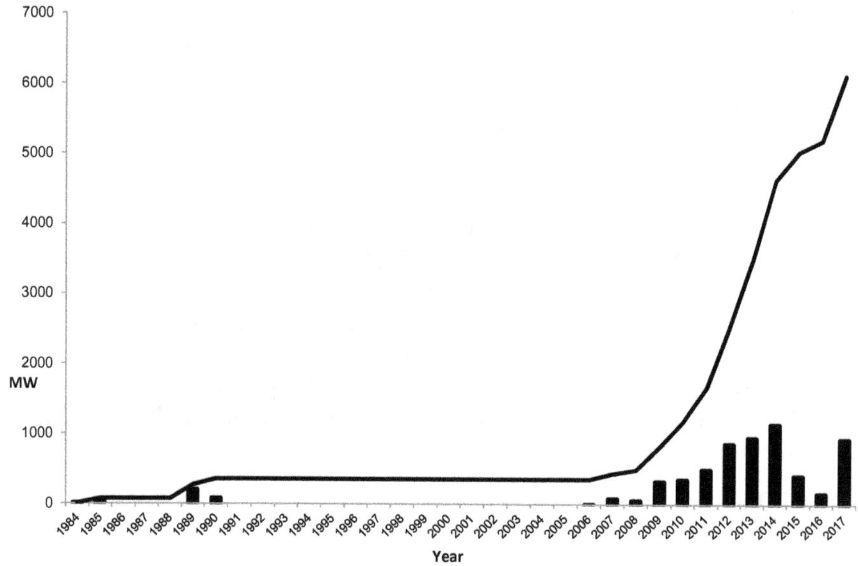

Fig. 3.1 Installed CSP capacity in the world (1984–2017)

This historical evolution has not been in line with the expectations. For example, in 2011, the world installed capacity in the middle of that decade was expected to be 2.6 GW [93: 25, 73], a figure which was substantially exceeded. There were other forecasts which are unlikely to meet the expectations: 30 GW of solar thermal electricity capacity was expected for Europe in 2020 by [49: 3712], although it seems that they will only be 2582 MW according to the trend between 2013 and 2016 [27: 22].[15] This amount is lower than the 6765 MW envisaged by the National Renewable Energy Action Plans of the EU Member States. Three scenarios which envisaged a minimum of 14 GW and a maximum of 40 GW in 2020 were considered in the study by Viebahn et al. [113: 18–20]. In addition, these numbers increase fast for the following decades. It is not clear that, with the values accumulated until 2017, even the lower of these two figures will be achieved. However, in CSPGO09 [16: 56] three scenarios for 2020 were also considered: a global capacity of 7.36 GW for the reference scenario, which would be 68.6 GW according to the moderate scenario and 84 GW for the advanced one. It all seems to indicate that the real number will be somehow above the reference scenario. As it can be observed, the expectations of the analysts change drastically in a few months. This is not surprising since the end of the previous decade experienced the end of the expansion cycle of CSP in Spain, with no countries clearly taking over, in addition to the sharp reduction of PV costs.

Since 2005, the annual installed capacity has been concentrated in a few countries, as shown in Fig. 3.2. As mentioned before, all commercial capacity was installed in the USA. According to Fig. 3.2, Spain (2008–2012) and the USA (before 2008

[15]This trend only applies to the EU.

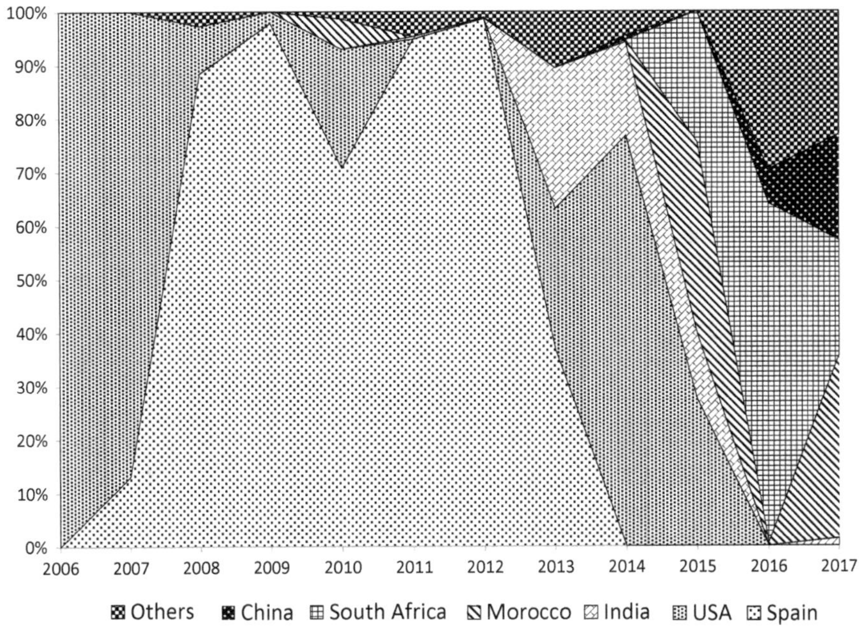

Fig. 3.2 Key countries in CSP world demand (2006–2017)

and in 2013–2014) stand out in this regard. Next, some countries emerged, such as India, Morocco, South Africa, and China, with more than 200 MW installed in the considered years. The group of other countries includes (in descending order with respect to the installed capacity) Israel (with 121 MW), UAE, Australia, Kuwait, Algeria, Saudi Arabia, Egypt, Mexico, Italy, Thailand, Germany, Canada, and Turkey (with barely 1 MW).

As of 2017, the distribution of the accumulated operative capacity per country is shown in Fig. 3.3. Spain still leads the ranking because many new projects in the USA are still not connected to the grid. However, the absence of new investments in Spain since 2013 suggests that USA will rank first shortly. This may happen with other countries as well (as suggested by the data in Table 3.1).

The figure may somehow be compared with the distribution of the installed capacity at the end of 2008: about 482 MW were distributed between the USA (419 MW, basically of the SEGS plants), Spain (63 MW), Australia (0.63 MW) and very small amounts in a few other countries [82: 43]. A few months later (mid-2009), the landscape was changing: There were 722 commercial MW (in operation or under construction). Experimental plants accounted for 367.5 MW (in the UAE), 332.3 MW (in Spain), and 22.25 MW (in other countries). In the beginning of 2011, the world CSP installed capacity was 1.26 GW, 58% of which were deployed in Spain. 500 MW was operational in USA and 1.5 GW was under construction. In the North of Africa and the Middle East, the plants in the pipeline amounted to 1.2 GW.

Table 3.1 Distribution of CSP technologies per country (March 2018)

MW	Trough	Tower	Fresnel	Dish	Total
USA	1708	2411	10	15	4144
Spain	2330.5	68.9	31.4	0	2430.8
Chile	370	625	50	0	1045
South Africa	350	250	130.58	0	730.58
India	406	2.5	100	1	509.5
Morocco	383	100	1	0	484
Israel	300	127.1	0	1	428.1
China	353.68	51	0	1	405.68
Egypt	120	250	0	0	370
Tunisia	50	205	0	0	255
Australia	0	106.5	80	40	226.5
Italy	130	50.35	0	0	180.35
Greece	0	50	0	75	125
Kuwait	110	0	0	0	110
UAE	100	0.1	0	0	100.1
Cyprus	0	25	0	50.76	75.76
Brazil	51	0	0	0	51
Jordan	50	0	0	0	50
Algeria	25	7	0	0	32
France	0	0	21.25	0	21.25
Argentina	20	0	0	0	20
Iran	17.25	0	0	0	17.25
Mexico	14	0	0	0	14
Thailand	5	0	0	0	5
Lebanon	2.8	0	0	0	2.8
Germany	0	1.5	0	0	1.5
Canada	1	0	0	0	1
South Korea	0	0.2	0	0	0.2
Total	6897.23	4331.15	424.23	183.76	11,836.37

Note Plants in operation, under construction, planned, and under development

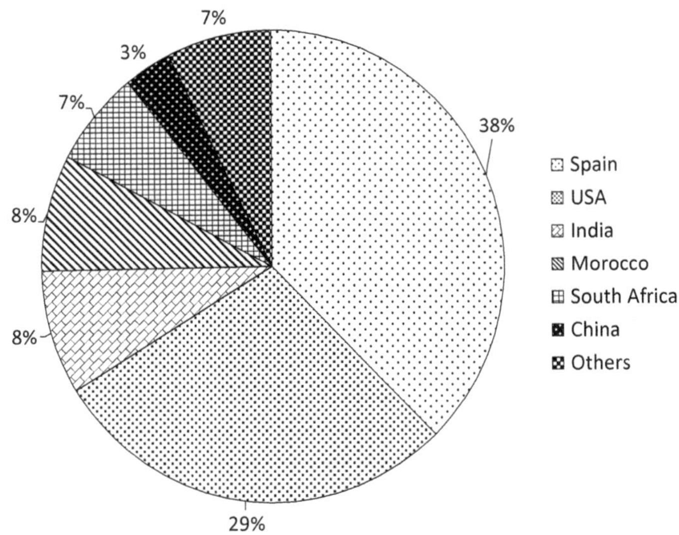

Fig. 3.3 CSP accumulated capacity per country (2017)

In April 2017, there were 96 commercial plants in operation, 22 under construction, 21 under development, and one which was not allocated [78]. Its distribution per technology is shown in Fig. 3.4. In 2009, parabolic trough collectors represented more than 90% of the installed capacity (673.25 MW), with 39 MW for solar towers, 1.5 MW for dish/Stirling and 8.4 MW for Linear Fresnel plants (data from [10: 2444]). Therefore, the dominance of the parabolic trough technology would be decreasing with respect to the solar tower. This trend will be maintained during the next years, as shown by the information provided below.

Table 3.1 shows the distribution of the CSP capacity which is either installed or expected, per country and technology, as of March 2018. All plants have been considered, both experimental and demonstration ones, as long as they feed at least part of their electricity to the grid. The table includes plants in operation, under construction, planned and under development. All of them have been included, except those which appear to be decommissioned or withdrawn. The plants whose goal is to produce steam for industrial uses, desalination, or oil recovery have been removed. Overall, 206 installations have been considered.[16]

Beside confirming the recovery of the USA leadership, the information displayed in the table shows a slight increasing trend of solar tower technology (36.6%) against

[16]The sources being consulted for Table 3.1 have been the database "CSP World Map" (see http://cspworld.org), "CSP Global Tracker" (see http://tracker.newenergyupdate.com/tracker/projects) and "CSP.guru" (see http://www.csp.guru). They have some gaps which we have tried to cover by searching information in Internet. Although these databases do not include declaration of intentions, some projects in the pipeline may never materialize. However, their impact on the numbers provided would be irrelevant.

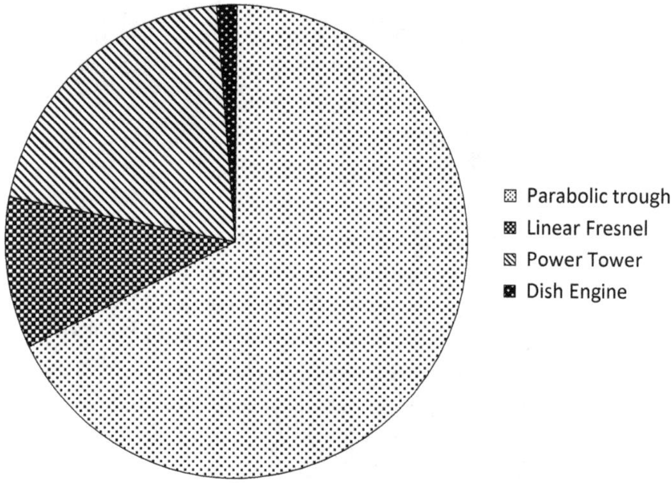

Fig. 3.4 Distribution of capacity per technology (2017)

Table 3.2 World CSP electricity generation in 2016 and expectations

Policies		Current policies[a]		New policies[a]		Sustainable development[a]	
Generation/year	2016	2025	2040	2025	2040	2025	2040
Generation (TWh)	6021	8840	13,160	9316	15,688	10,625	22,664
CSP (%)	0.18	0.41	0.98	0.47	1.51	0.93	4.70
PV (%)	5.03	12.39	16.65	13.56	20.15	15.33	23.32
Wind (%)	16.29	22.43	25.51	23.52	27.21	26.21	30.66

[a]According to OECD/IEA [80: 37–38], the Current Policies Scenario means that countries implement only policies and measures enacted by mid-2017, the New Policies Scenario mixes existing policies and announced policy intentions, and the Sustainable Development Scenario refers to the goals of the "2030 Agenda for Sustainable Development" adopted in 2015 by the UN member states

parabolic trough (whose share progressively decreases, but it still dominates with a 58.3%), whereas the other two technologies barely add 5%.

Another issue to consider is the current and expected share of solar thermal electricity in total generation, whether from a renewable or conventional origin. Table 3.2 shows the current and expected values according to different scenarios (own elaboration from OECD/IEA [80: 299]).

In Table 3.2, the share of CSP is always very modest, and it does not reach 5% of all the electricity generated, not even in the most favorable scenarios for renewables. When compared to PV or wind generation, the share of solar thermoelectricity is between 5 and 20 times lower.

In a nutshell, the recent evolution and current situation of the sector can be summarized in a few sentences: CSP generation is mainly made up of parabolic trough plants, whose precedent were the Californian installations deployed during the 1980s and which were the only ones operative in the world for more than 15 years, next the sector revived in Spain but only for a short period of time (2007–2012) and, finally, despite the fact that total installed capacity has not increased in a sustained manner, the deployment of new installed capacity has emerged in other countries with Mediterranean climate and/or arid zones.

3.2.1 The Installed Capacity Stages from 2007 Onwards

More than ten years have passed between the resurgence of the sector in the 2000s and today. In this short time span, two main phases can be identified: (i) from 2007 to 2012, when all new installed CSP capacity was deployed in Spain and the colossal Desertec project was being advocated for and (ii) the period from 2013 until now.

The first stage coincides with the boom of renewables in Spain, whose peak was reached in 2007–2008. Nevertheless, for the case of solar thermal electricity, deployment materialized in the subsequent years due to the long lead time that runs from the elaboration of the project, the search for financial resources and the construction and test of the plant. More specifically, a boom of projects and intentions occurred in 2009, and their progressive accomplishment extended until 2013 [67, 73: 450–465]. According to the annual share shown in Fig. 3.5, a total of 2430 MW were installed.

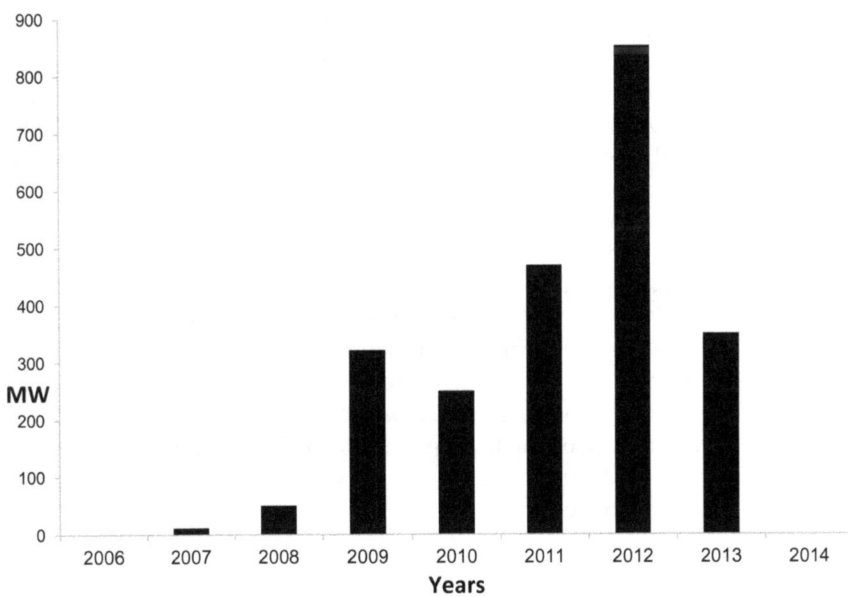

Fig. 3.5 Annual CSP capacity installed in Spain (2007–2014)

After a successful start of the wind energy sector in the 1990s, the Spanish governments decided to boost solar electricity generation, both PV and solar thermal (for a more detailed description of CSP regulation see Chap. 6). The most significant milestones started in 2004, a year when a favorable remuneration scheme for renewable energy sources was adopted. This scheme was later improved in 2007. As a result, there was a solar thermal boom. However, after the experience of the PV boom in 2007/2008, the authorities and the CSP promoters achieved an agreement in November 2009 to freeze the expansion of installed capacity. For this reason, the initially announced 19 GW by 2020 were finally reduced to 2.4 GW. All in all, Spain became the world leader in CSP and has maintained its prominent position until today.

The aforementioned regulation in 2004 (Royal Decree 436/2004) set the basis of the regulation for solar thermal electricity. For example, the hybridization with gas to increase the HTF temperature was allowed up to a given percentage of the MWh generated. Three years later, Royal Decree 661/2007 expanded the list of hybridization fuels, increasing the share of annual production obtained by burning biomass or biogas. Regarding remuneration, the possibility to either receive a given preferential tariff (updated periodically) or to sell the electricity in the wholesale electricity market, adding a premium to its price, was offered (no distinction among CSP technologies was made). Obviously, there were upper and lower limits to the remuneration: the producer would receive at least the so-called reference premium. The premium would go down if the price of electricity would go up. If it exceeded a given upper limit, then the companies would not receive the premium (see [73: 450–461] for further details).

At the very beginning of the considered period (2007–2012), nothing significant happened in the CSP sector in Spain. Leaving aside experimental installations developed for years in the Plataforma Solar de Almería, the commercial projects came in dribs and drabs. However, in 2007 the solar tower PS10 started operating in Sanlúcar la Mayor (Sevilla, Spain), with a capacity of 11 MW. The solar thermoelectric sector started to make the news in 2008 and reached its zenith in 2009, when large construction and engineering firms declared that they were developing commercial projects.[17] Thus, whereas at the start of 2009 there were six solar thermal electric plants in operation, another 27 were in the pipeline, which represented an accumulated total of 1037 MW. This figure would have increased to 80 projects (4 GW) at the end of 2009 (and many more if we consider the announced intentions day after day). These figures were much higher than the 500 MW envisaged by the regulator by 2010. This acceleration of the sector, which suggests that a boom was being formed, led the authorities to establish a registry of projects in order to administratively control the pace of expansion, as well as to increase the guarantees that the project developers had to provide.

[17] Solar thermal generation had become an interesting investment in itself and as a way to diversify assets. For this, these companies had huge funds which had been accumulated during the Spanish economic boom (1996–2008).

The agreement between the CSP sector and the authorities in November 2009 allowed 2400 MW of additional thermo-solar capacity. More specifically, 860 MW were expected for 2010, 500 MW per year in 2011 and 2012, and the rest in 2013. In total, there was room for around 55 new plants. With respect to the retribution, Royal Decree 1614/2010 limited the number of generation hours which could be eligible for the premium, distinguishing between technologies and generation modalities. This cut the total remuneration, but also protected the installed capacity from greater tariff reductions. Therefore, the agreement set a guarantee of authorization for the most mature projects and the right for an attractive remuneration for the useful lifetime of the plant, but this was swapped by a limit on the number of hours with a right to receive the premium and a reduction in the number of MW which could be installed.

In the beginning of 2012, a new government approved a moratorium on new renewable energy capacity. Two and a half years later, the production-based remuneration scheme which was in force until then was modified by a capacity-based remuneration scheme whose values could be readjusted every three years, without a systematic criterion being provided for the medium and long terms. The different regulatory reforms undertaken since 2010, which declined the remuneration of all the plants in operation (see Chap. 7), led to a serious judicial conflict with respect to retroactivity [22, 32, 75]. In 2016, solar thermal electricity was not eligible to participate in the call for auctions for the allocation of new renewable energy capacity. In the two auctions held in 2017, no CSP projects were awarded (CSP was eligible to participate in the first one of these two auctions, the one taking place in May, but not in the one organized in July).

During the same period, the Mediterranean Solar Plan (MSP) was born. Announced on July 13, 2008, at the Paris Summit for the Mediterranean Region and formulated at the end of this year, the MSP aimed to deploy an additional 20 GW of renewable electricity generation (half from CSP) in the Southern and Eastern Mediterranean Basin by 2020 [16: 69, 48]. The initiative was registered in the Union for the Mediterranean, a forum for the political-economic collaboration between countries of the Mediterranean rim which had been proposed some months before by the then French president Nicolas Sarkozy, who was also the President of the European Council. As the first task, the MSP team elaborated an inventory of all the renewable energy projects existing in the Mediterranean basin, with the exception of those of the European Union Member States. 150 initiatives were catalogued, 3/4 were solar projects and the rest were wind farms [65: 16–17]. Additionally, the MSP received funds from the EU in order to carry out several studies of EU/MENA energy cooperation with a perspective on the medium and long terms. It was assumed that the countries with more prominence in this plan were Germany, France, Spain, Italy, Egypt, and Morocco [58, 65: 19]. The master lines of the MSP were not approved, however, until the middle of 2012 in the framework of the Mediterranean Energy Forum held in Brussels.

Also during the same period, the Desertec Foundation was established on January 2009 (see www.desertec.org). This NGO sponsored the studies on a possible energy future based on renewable energy plants located in MENA countries, with the aim to supply energy to them as well as Europe. Among the institutions in charge of developing this vision, the reports of the German Aerospace Center (Deutsches Zentrum für Luft- und Raumfahrt e.V., abbreviated DLR), a center for aerospace, energy and transportation research whose headquarters are located in Cologne, stand out. In their reports, the DLR considered all the renewable energy sources (see, for example, [110]). The best expectations were for wind generation (with possibilities in some places in the North of the Mediterranean and Turkey), PV (Sahara, Arabia desert and places in Syria, Iraq, and Iran) and, of course, solar thermal electricity. Without this one, the available resources for renewable electricity generation were estimated at 1940 TWh/year. However, if CSP generation was added, such figure would increase 300-fold. The volume of thousands of TWh/year would be several orders of magnitude higher than the electricity needs of the whole planet [104: 56]. In fact, it was estimated that transforming only 0.04% of the incident solar energy in the Sahara Desert, the electricity demand of Europe could be met, whereas 2% of its surface would be enough to meet the electricity demand of the world [16: 69]. It was expected that, in 2050, and considering the best sites and technical improvements, wind, hydro, biomass, geothermal and PV would provide between 200 and 300 TWh/year each. Regarding solar thermal electricity, it could provide 2200 TWh/year, that is, as much as the rest of the renewable energy technologies together. Still, this figure would barely represent 1% of its maximum capacity [104: 13]. Before such date was reached, it was estimated that solar electricity generation could supply between 20/40 TWh/year in 2025 and between 120/140 TWh/year in 2035. The capacity would be allocated through auctions and the financing of those projects should be guaranteed through long-term PPAs together with (if necessary) grants and subsidies [107: 316]. All these forecasts were compared to the 3500 TWh/year of electricity which was consumed by European countries in those years, a figure that the MENA countries could reach in 2050.

Inspired by the ideas developed by the Desertec Foundation, the consortium Desertec Industrial Initiative GmbH was also created in 2009. It brought together several large industrial firms of the energy and the financial sectors from Germany (Siemens, E.ON, Schott Solar, Deutsche Bank, Munich Re, RWE, etc.), Spain (Abengoa Solar) and other countries [14, 110]. In July of that year, the consortium announced its willingness to invest €400 billion ($510 billion) in the next four decades to achieve the goal that 20% of the European electricity demand would be met with exports from the MENA countries in 2050. This was a huge figure, which would benefit from the large spaces of land which is available without a permanent population in those countries and a high DNI. For example, DNI levels above 5 kWh/m^2/day can be observed in Morocco (except in the Atlas), Algeria, Tunisia, Libya, and Egypt (except the coast), the Arabian Peninsula, a large part of Israel, Jordan and Iraq, the South of Syria and the islands of Cyprus and Crete [111: 8, 119].

More specifically and in the context of this book, taking into account the level of irradiation, the type and inclination of the field and access to road infrastructures and gas fields, the best places to erect solar thermoelectric plants would be the East of Morocco, the provinces of Illizi and north of Tamanrasset in Argelia, the regions of Nalut, Al Jabal al Gharbi and Al Jufrah in Libya, the south of Tunisia, the center and south of Egypt, some places in the East of Jordan and the triangle Mecca/Riyadh/Ha'il in Saudi Arabia. In total, these outstanding locations would add up to more than 110 thousand km^2, 80% of which would be in Libya and Saudi Arabia [111: 13]. In this area, more than 538,000 TWh/year of solar electricity could be generated. However, it would be enough for the plants to occupy 0.2% of such surface in order to provide 15% of the European electricity demand expected in 2050 [108: 341].

A major obstacle that the Desertec project faced was the connection of the electricity grids of the involved countries. The electricity interconnections were weak between both sides of the Mediterranean, as well as between neighboring countries. The different studies carried out [71, 104, 108: 346–352, 111, 112: 313] indicated that undersea cables should be installed in the western Mediterranean sea (through Gibraltar, from Tunisia to Sicily and the Italian peninsula or to Sardinia, Corsica, and Liguria) and between Turkey and the European continent, in addition to interconnections between neighboring countries, with the aim to increase the export capacity to 7 GW in 2020. Overall, in 2050, 33 HVDC trans-Mediterranean power lines, with a nominal capacity of 4 GW each, would be required to transport 703 TWh/year of electricity from solar origin. This is a huge infrastructure whose construction would take three decades, with an investment amounting to €666.5 billion (in constant monetary value of 2010) [108: 343]. If the expenditures in the rest of lines were added, the total investment in grids would not be lower than €3100 million.

The willingness of these countries to export electricity to Europe in the framework of the Trans-Mediterranean Renewable Energy Cooperation was also highlighted [104: 35, 106, 108].[18] The initial investment to deploy renewable energy technologies through the MENA region was estimated at about $75 billion until 2020. For 2050, the total amount to be invested would be close to $250 billion [104: 157].[19] These huge numbers, however, came along with optimistic expectations with respect to generation costs. Table 3.3 shows the values of the total investment, in $/kW, and the generation cost, in ¢$/kWh, for the most relevant technologies between 2010 and 2050 (data from Trieb et al. [104: 128]).[20]

As it can be observed, the cost of the kWh for CSP in 2020 was expected to be lower than the costs of PV and conventional fossil fuel technologies, although slightly higher than the costs for wind. It was expected that these lower costs would be maintained until 2050.

---

[18] However, there were concerns about the threat to energy security in Europe due to the dependence with respect to third countries. See Lilliestam and Ellenback [59] and Lilliestam et al. [61] for a detailed analysis of this issue.

[19] Those investments should be part of a global industrial development plan in the Southern shore of the Mediterranean [55, 117].

[20] According to the table located in page 127 of this source, the economic life (years), efficiency level (%), fuel price escalation (%), O&M cost by % of investment/year, and annual full load hours considered in calculating the values on Table 3.3 were 15, n.a., –, 1.5% and 2000 (for wind power); 20, 10%, –, 1.5% and 1800 (for PV); 40, 37, 1, 3% and 8000 (for CSP), and 30, 40–48, 1, 2.5% and 5000 (for gas/oil), respectively. For more information, go to the source of the table. Moreover, the costs of transport should be added. These costs are around €cents 1/kWh (see Trieb et al. [105: 4]): Thus, the Algerian generated kWh would cost between €cents 4.3–5.5/kWh in Germany by 2050 [114: 4426].

Table 3.3 MENA costs of generation

Year/Tech.	2010		2020		2030		2040		2050	
	$/kW	¢$kWh	$/kW	¢$kWh	$/kW	¢$kWh	$/kW	¢$kWh	$/kW	¢$kWh
Wind	1280	5.2	950	3.8	930	3.7	920	3.7	900	3.6
PV	2830	14.7	1590	7.4	1250	5.8	1010	4.6	910	4.2
CSP	3388	7.1	4662	5.2	4332	4.5	4185	4.1	4134	4
Gas/oil	540	5.8	530	6.2	520	6.7	510	7.2	500	7.8

Other sources estimated similar values. For Ummel and Wheeler [111: 19], the cost of generating a kWh of solar thermoelectricity would be between 12.8 and 16.7 ¢/kWh, which could be ¢6/kWh–¢12/kWh by the end of the decade [82: 45] or €cents 7/kWh in 2030 (according to [112: 332–334]).

Regarding some facts of the Desertec Industrial Initiative, the cooperation agreement reached with the MSP in May 2012, with the aim to coordinate efforts for the development of electricity generation projects in the Mediterranean region, should be mentioned. The memorandum of understanding signed, at the end of 2012, between government representatives of several EU countries (with the exception of Spain) and Morocco was a major milestone for the interests of Desertec. In this agreement, the German firm RWE was commissioned to assess the development of a CSP plant of 150 MW, together with a PV plant of 100 MW, and a wind farm of 100 MW, whose electricity would be brought to Europe through the Italian and Iberian peninsulas. It should be taken into account that, some months before, Morocco had organized a site-specific auction with a volume of 160 MW of CSP generation to be deployed in the south of the country (Ouarzazate region). This was the first phase of a 500 MW project financed by the World Bank and the African Development Bank, as well as other European institutions, including the European Investment Bank (EIB), the Development Agency for France (AFD), and Germany's KfW Entwicklungsbank (KfW).

Some weeks later, the Desertec Industrial Initiative signed an agreement with Egypt for the deployment of renewable energy projects in the country (without specifying the details). Around the same time, the State Grid Corporation of China and the US PV module producer First Solar Inc were interested in participating in the Desertec project. However, in early 2013, Siemens and the Bosch subsidiary announced their intention to abandon Desertec, alleging internal economic reasons. At the time, the bad health of the consortium was obvious and, in March 2013, their managers declared that they gave up the objective for 2050. There are many reasons that partially explain the failure of this initiative [61]. Among others, the political instability of many MENA countries was a compelling factor. It should not be forgotten that, although the European Commission had accepted to partially finance a feasibility study on the connection of the Italian power grid to the North African grid, it showed doubts at the end of that year as to whether to get deeply involved in the financing of this megaproject [109].[21]

At the end of 2014, the Desertec Industrial Initiative announced a change in its corporate strategy: rather than promoting specific industrial projects in order to achieve the 2050 objective, it would be a provider of services, primarily to its shareholders, for the construction of renewable energy projects in the MENA region.

The second phase starts in 2013 with the end of the Spanish preeminence and the recovery of the activity in the USA, but without the relevance it had in the 1980s and 1990s due to the arrival of new actors, such as Chile, China, Morocco, South Africa, and other sunbelt countries. In these last years, the remuneration through administratively set FITs has been abandoned (with the notable exception of China), and there is a trend toward allocating capacity through auctions, whose price would lead to a long-term PPA. This trend has occurred for all renewable energy technologies [43], including CSP [101].

After 2014, the market growth rate was slowed down and it was not until 2018 onwards that it started to recover again. In 2016, the operative accumulated capacity was 4889 MW. Its distribution and prospects were as follows [27]:

- Almost half of the capacity was concentrated in Europe (2313.7 MW, including pilot plants). However, the prospects were poor. In addition, numerous lawsuits had been promoted against Spain by the investors who had been seriously affected by the severe cuts to the remuneration since 2010. Thus, it was estimated that the revenues of CSP plants had been reduced by a third.
- North and South America had 1758 MW.
- Middle East emerged with 123 MW. This was a modest figure, but with good prospects: Saudi Arabia was building Duba 1 (a parabolic trough project of 43 MW) and the Waad Al Shamal plant; Israel was building the Ashalim power station, a solar tower of 250 m high; a program of 1 GW CSP was envisaged in Dubai, etc.
- In that year, 260 MW were added in Africa, which increased its accumulated capacity up to 429 MW. It was the most active continent, with the start of operation of Noor 1 (392 MW) in Ouarzazate (Morocco), whereas Noor 2 (parabolic trough) and Noor 3 (solar tower) (plus 70 MW PV) are expected. Adding them all, the overall project capacity will reach 580 MW. The other leading country was South Africa, whose solar tower Khi Solar One, of about 50 MW, started operation at the beginning of 2016. A parabolic trough plant of 50 MW was under construction in Groblershoop, together with Xina Solar One, of 100 MW and Kathu Solar Park, of 100 MW, both parabolic trough; and the Solar Tower of Redstone, of 100 MW, in total, 400 MW.
- Barely, 10 MW were connected in Asia, with a total capacity of 268 MW. In China, which had around 20 pilot plants (with a total capacity of 1.4 GW), the parabolic trough plant SunCan Dunhuang started to operate at the end of 2016. A 100 MW solar tower was also expected.

[21] Around those dates, the collaboration between the Desertec Industrial Initiative and the Desertec Foundation came to an end, although the later would continue developing the vision of the project.

- The installed capacity in Australia and Oceania was 6 MW (with half of this capacity installed in 2016).

As mentioned above, some firms from the sector suffered several important crises, as it was the case with Abengoa. Solar Millennium abandoned the sector, whereas Schott and Rioglass merged in 2016. The Spanish firm Ibereolica, which was one of the major CSP project developers, went bankrupt in 2016. Chinese firms also emerged, however (see Chap. 4).

3.2.2 Evolution of Generation Cost

The availability of data to assess the historical trend of generation costs of CSP plants is very limited. Scarce and disperse data are provided by the different sources. Fortunately, the developers of the database www.csp.guru [62] have made a remarkable effort to concentrate as much information as possible on CSP plants and to harmonize the economic information on them. This will be the main source in the following pages since it is currently the best database, in terms of exhaustivity and considered variables.

To start with, Fig. 3.6 shows the evolution of the LCOE, estimated for a WACC of 5% and expressed both in € (gray line) as well as in $ (white line), from 1984 until 2018.[22] All the data are in 2017 monetary units and a differentiation between the distinct technologies has not been made. It should be mentioned that the data for 1998 comes from Price and Kearny [89: 12] and has been deflated and converted to euros. As expected, the figure shows many years without any data since there weren't any commercial plants being deployed.

Expressed in €, the first LCOE value (1984) is €cents 78.8/kWh and the last one (2018) €cents 11.09/kWh (¢66.2 and ¢11.1/kWh, respectively). Therefore, the estimated LCOE would have been reduced sevenfold (€) or sixfold ($). It can thus be concluded that a substantial reduction of generation costs has taken place. A closer look, however, shows comparatively higher values between 2008 and 2013, followed by a reduction since 2015 onwards. There are four reasons which explain this evolution [26: 833–842, 41: 30–31, 44, 62: 46, 80, 84–85]:

- In those years, virtually all solar thermal plants were deployed in Spain, given the generous FIT being provided. The country, however, is in the upper range (40° North) of the climate strip, with the best DNI levels for solar thermal generation.
- Since 2012, the projects have spread over places such as Morocco, North of Chile, Southwest of the USA, and UAE, all of them with high DNI levels.
- The reactivation of the sector expectations has facilitated securing financial resources at more favorable conditions.

[22]The database includes values up to 2020. Only plants in operation or closer to operation have been considered.

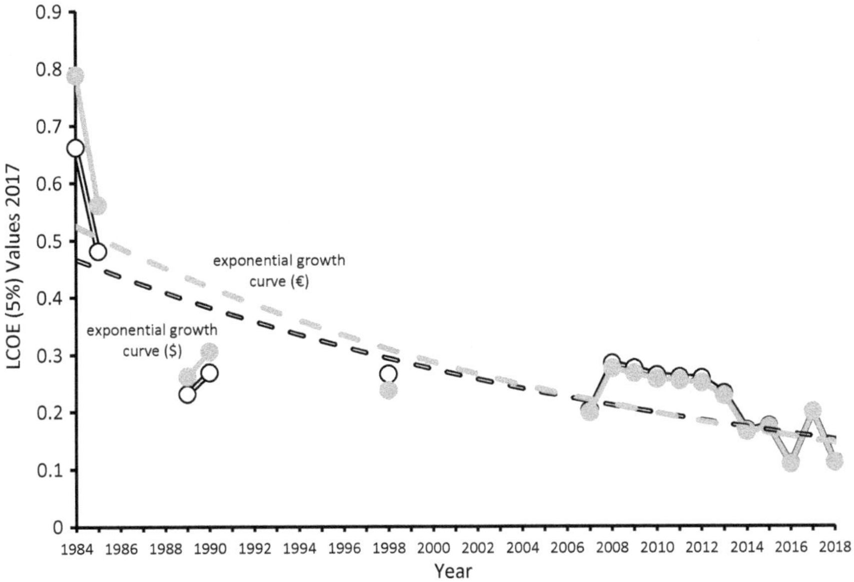

Fig. 3.6 LCOE of CSP generation

- The experiences of SEGS, the commercial plants in Spain, and the experimental installations in many places have enabled the introduction of improvements in components and systems.

Figure 3.6 also includes an attempt to fit the data to an exponential growth curve. The aim of this exercise is to do a mere extrapolation to suggest possible future values.[23] The following results have been obtained:

- For the values expressed in €, the first fitting value is €cents 52.48 and €cents 14.58/kWh is the last value. The extrapolation which has been carried out indicates an LCOE of 11.2/kWh in 2025.
- For the values expressed in $, the initial fitting value is ¢46.63/kWh, and ¢15.13/kWh is the last value. The extrapolation which has been carried out indicates an LCOE of ¢11.9/kWh in 2025.

The projected values reflect the improvement in the expectations of CSP cost reduction resulting from the reactivation of the sector in the last few years. This can be contrasted with relatively recent studies (such as, for example, Kost et al. [56: 31–32]):

- This study reviews several projections undertaken in the end of the last decade which, for 2025, suggest a LCOE (for a plant whose investment is €6000/kW and a DNI of 2500 kWh/m^2/year) between €0.115 and €0.132/kWh. These values are closer to the ones obtained from the extrapolation exercise.

[23]The time series tool of the Minitab program has been used.

- On the other hand, the same study predicts a LCOE for 2030 between €~0.09 and €~0.115/kWh (for a learning rate of 10%, DNI from 2000 to 2500 kWh/m^2/year and average market development). The values obtained in our extrapolation are very similar $0.1016/kWh and €0.0927/kWh.

Other studies have suggested comparable values. This is the case of Hernández-Moro and Martínez-Duart [37: 191]. Starting from an initial investment (for 2010) of $4200/kW to $8700/kW, and a cost of generation between ¢12.5 and ¢22.5, they envisage a LCOE of ¢11/kWh, ¢9.5 and ¢8/kWh, in 2020, 2030, and 2050, respectively.[24] More recently, however, a weighted average cost of $\$_{2016}$0.072/kWh has been forecasted for 2022 [44: 86, 45]. It should be noted that this value is slightly below our extrapolation figure. Nonetheless, as these sources warn, this expectation relies on auction prices, which are not the same as LCOE. Moreover, this forecast can only be applied to projects which will be commissioned in the period 2020–2022. Finally, there is no doubt that projects with particular advantages and features can bid very low values, as shown by the results of recent auctions in Australia and UAE in which winners bid with ¢6 and ¢7.3/kWh, respectively, (see Chap. 6).

Figure 3.7 has been built combining a forecast of the costs per kWh for parabolic trough plants published in 1999 and real LCOE data from 2007 to 2016.[25] As it can be observed, the proposed extrapolation estimates a LCOE of ¢4.4/kWh ($1998) by 2020. If the estimated trend was accomplished, the prediction made 20 years earlier would have been confirmed. However, the trajectory would have been very different to the one imagined before: 15 years without any activity would have involved a new start, although the later reduction in generation costs would have allowed to make up for the lost ground.

Another study published at the start of this century which stands out for its projections of both parabolic trough and solar tower technologies' LCOE is SL [100]. In this study [100: ES-4/ES-13], the high-cost bound of the kWh generated by a parabolic trough plant in 2020 was estimated at ¢6.2, whereas it was ¢5.5 for a solar tower. Between 2002 and 2020, a cumulative deployment between 2.8 GW (parabolic trough) and 2.6 GW (solar tower) was assumed. The authors of the report compared their predictions with those of another study carried out by SunLab and the indus-

---

[24] These numbers were calculated on the basis of the following variables: a DNI of 2850 Kwh/m^2/year, a learning ratio of 10%, annual O&M costs of 2% of total system costs, a useful lifetime of the plant of 30 years, an annual output reduction of 0.2% due to the degradation of the turbines, a 10% real discount rate, and a performance factor of 0.853 m^2/kW. An accumulated installed capacity of 1100 GW in 2050 is assumed, according to the CSP Roadmap scenario of the IEA [38].

[25] More specifically, on the one hand, the initial hypothetical value in the 1999 Roadmap for 2000 was ¢14/kWh, which was assumed to increase to ¢4/kWh in 2010 [89: 17]. On the other hand, the average LCOE annual values (according to a WACC of 5%) for 2007–2016 have been calculated according to the data included in www.csp.guru, although the 1998 number comes from Price and Kearney [89: 12]. All these values are expressed in $1998, according to the CPI series published in https://data.oecd.org/price/inflation-cpi.htm, and assuming an annual inflation of 2% for 2017 and 2018. The line reflects the linear trend for 2008–2018, which expression is $Y_t = 0.2312–0.0144t$, together with a simple extrapolation for 2019 and 2020. This was calculated using the Minitab program.

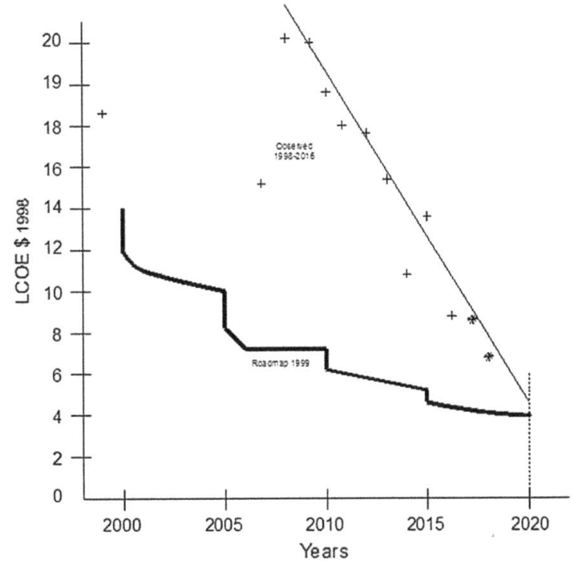

Fig. 3.7 An early forecast

try. This study was much more optimistic: in 2020, the LCOE would be ¢4.3 and ¢3.5/kWh for parabolic trough and solar tower plants, respectively, with a cumulative installed capacity of 4.9 and 8.7 GW.[26] It should be noted that both predictions are not unreasonable and seem to have been fulfilled in general terms, although none of them considered the issue of the evacuation of the electricity generated.

Another aspect to be taken into account when assessing the costs of CSP is their considerable dispersion. Figure 3.8 shows the LCOE values (WACC 5%, €-gray- and $-white-) for the years in which, according to the data available in www.csp.guru (on May 2018), the plants started to operate.

Figure 3.8 shows that, in the years 2011, 2012, 2013, 2014, and 2018, when many plants entered into operation, the maximum estimated LCOE was more than twice the minimum value. This fact depends on many factors, namely

- The DNI level: the available data suggest that the capacity-weighted average DNI value from 2014 onwards was among ~2400 to ~2800 kWh/m^2/year, versus ~2000 and ~2300 kWh/m^2/year between 2009 and 2013 [44: 83–84]. This fact is inversely correlated with the generation costs.
- The existence of a storage system: the trend to design plants with a growing number of storage hours has a direct impact on the investment and, in some cases, on the LCOE. Thus, a plant with four hours of storage has an installed cost of $6050/kW, which increases up to $12,600/kW for 8 h. However, the capacity factors of plants without storage are usually below 30% and above 50% for plants with more than 8 h of storage [44: 81, 84].

[26] All these monetary values are assumed to be in current terms for the corresponding year. The economic assumptions used are indicated in the original document.

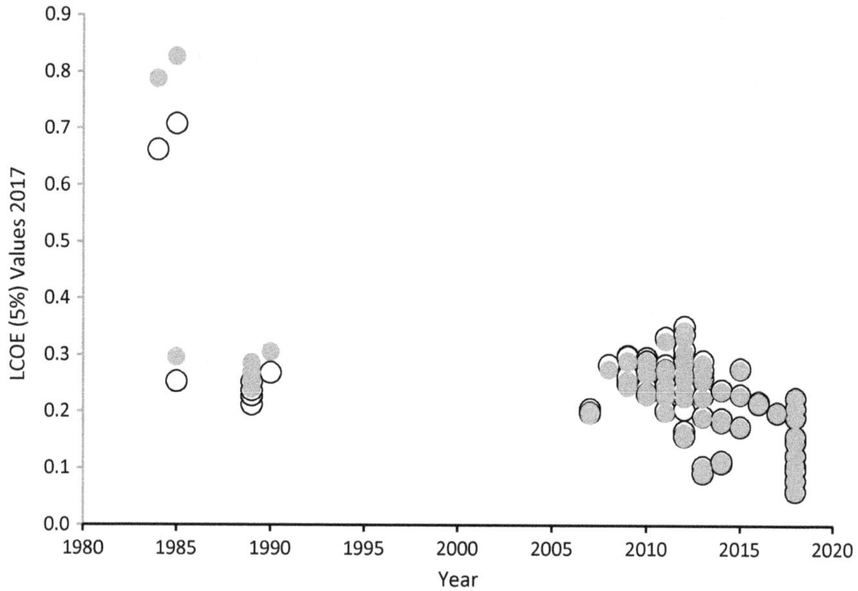

Fig. 3.8 Dispersion of LCOE

Table 3.4 Lowest LCOE of CSP in the considered year

Year	LCOE (5%) (€/kWh and $/kWh)
2007	Nevada Solar One (€0.1374, $0.1898)
2008	Andasol (€0.181, $0.2707)
2009	Solnova 4 (€0.1693, $0.2401)
2010	La Florida (€0.1654, $0.2223)
2011	Arcosol 50 and Termesol 50 (€0.1375, $0.1936)
2012	Aste 1b (€0.1111, $0.1439)
2013	Supcon Phase 1 (€0.0663, $0.0886)
2014	Dhursar (€0.0795, $0.1057)
2015	Crescent Dunes (€0.1523, $0.169)
2016	Shenzen HuaQing Hebei (€0.1151, $0.1277)

- Other factors, including the financial conditions, the maturity of the supply chain, etc.

Based on the guru database, Table 3.4 indicates the names of the plants with the lowest LCOE in Fig. 3.8.

To sum up, the values shown in the previous paragraphs suggest that the generation costs of CSP depend on factors related to the design of the plant as well as on the conditions of the place where it is located.

3.2.3 Cost Reductions: Experience and Expectations

The learning rate (LR), which refers to the reduction in renewable generation costs for every doubling of installed capacity, is a complex issue. In addition to its methodological and theoretical problems (see [74: 211–219]), a wide empirical base is required. Unfortunately, differently from wind or PV generation, the large-scale deployment of CSP is just starting. Regardless, some authors have tried to calculate the learning rate. For example, Lilliestam et al. [62] estimate a learning rate above 20%. This value is twice as high the calculation of the experts at the end of the first decade [44: 83]. Indeed, the estimation of the learning rates for the main components of CSP plants carried out by Viebahn et al. [113: 31–32] indicated that, in the case of the power block (which is a very mature technology because it is also used in other thermal plants), the learning rate was barely 5%, whereas the LR of a solar field with TES was around 12%. Notwithstanding, although other authors suggested similar values, such as the LR of 10–12% for the whole CSP generation suggested in Hernández-Moro and Martínez-Duart [37: 186] or in Kost et al. [56: 31–32], other studies such as De la Tour et al. [19: 9] were more optimistic by considering a LR of 17%. The non-negligible magnitude of those differences reflect the limited nature of the empirical base on which they were sustained and the assumptions made, as well as the different perceptions on the expectations of the sector. This last factor explains that a LR of 30% has been recently estimated [44: 83].

In these pages, rather than providing a detailed account CSP LR literature review or proposing a new estimation, a comparison of the efficiency and investment costs of the main components and systems is provided, considering the past and current published data. This is complemented with some expectations in this respect.

The evolution of two main aspects determines the evolution of the generation costs of a CSP plant:

- The efficiency of the main components and systems of the plants, where the performance of the solar field, the storage process, and the turbine/generator stand out. Its dynamics depends on the improvements introduced in RD&D.
- The investment volumen per installed kW. In CSP plants, the costs of the collectors, supporting structures, TES, and the power block stand out. In turn, the costs of these systems depend on several factors: the scale-up of their manufacturing, technical improvements embedded in manufacturing (equipment, organization of tasks, etc.) and the size of the plant (i.e., its capacity in MW).

It should not be forgotten that, although the solar collectors/reflectors are to some extent modular, and, thus, their improvement and cost reductions can take place isolated from the rest of plant systems; this is not the case with solar towers and with the power block. SL [100] shows the large amount of information regarding efficiency and the investment costs of parabolic trough and solar tower plants at the beginning of this century. The authors, on the basis of previous studies, the most updated information collected at the moment and some economic models, projected the evolution of many technical and economic variables of CSP plants up to 2020,

Table 3.5 Efficiency and capital cost of parabolic trough power plants

Variables/type of plant/years		Base line	S and L	Recent data
		SEGS VI	Trough	SAM[d]
Characteristics	Year in service	1989	2004	2018
	Power (MW)	30	100	100
	Capacity factor	22% (only solar)	53.5%	39.3%
	Solar aperture area (m^2)	188,000	1138,709	877,580
	HTF operating temperature (°C)	391	391	391
	Thermal storage (hours)	No	12	6
Efficiency	Solar field optical efficiency	53.3%	56.7%	76.7%
	Receiver efficiency	72.9%	84.3%	96%
	Thermal storage efficiency	–	99.1%	98.5%
	Gross turbine cycle efficiency	35%	37%	37.7%
	Power plant availability	98%	94%	99%
	Solar-to-electric efficiency	10.6%	14%	20%
Capital cost[a]	Solar collector ($/$m^2$)	250	234[b]	150
	Power block ($/kW)	527	306[b]	910
	Thermal storage ($/$kW_{th}$)	–	958[b]	62
	Total plant cost ($/kW)	3008	4816[b, c]	5272

Source Own elaboration from S&L [100: 4–5/4–6, 4–11 and 4–28/4–29] and SAM [98]
[a]Current economic values
[b]Parabolic trough pilot plant of 50 MW
[c]Without storage, the total costs are 2453 $/kW
[d]Plant type: CSP parabolic trough (empirical), Commercial (distributed), Station ID 67,345 (Tucson)

taking the SEGS plants as a reference. In this work, the authors do not pay attention to their projections, but rather compare their base values with more recent ones from other sources. Table 3.5 includes indicators of efficiency and capital costs of parabolic trough plants, as well as some of its general features.

Data in Table 3.5 are merely indicative. The figures presented in the two first columns come from a report on the general situation of the sector [100], whereas the figures in the third column are default input performance and economic values of the last version (2018/11/11) of the well-known System Advisor Model (SAM). There is no doubt about the enormous case dispersion, whether due to plant characteristics or geographical location. However, the table shows improvements in solar field optical,

Table 3.6 Efficiency and capital cost of solar tower plants

Variables/type of plant/years		Base line	S and L	Recent data
		Solar two	Solar *Tres*	SAM[b]
Characteristics	Year in service	1996	2004	2018
	Power (MW)	10	13.65	115
	Capacity factor	21%	78%	63.7%
	Heliostat size (m^2)	48	95	144.37
	Solar field area (m^2)	80,000	244,966	1269,055
	Receiver area (m^2)	100	280	244.7
	Operating temperature (°C)	565	565	574
	Thermal storage fluid	Molten salt	Molten salt	Molten salt
	Thermal storage (hours)	3	16	10
Efficiency	Collector efficiency	50.3%	56%	51.5%
	Receiver efficiency	76%	78.3%	94%
	Thermal storage efficiency	97%	98.3%	99%
	Power plant availability	90%	92%	99%
	Solar-to-electric efficiency	7.9%	13%	22%
Capital cost[a]	Heliostats ($/$m^2$)	–	160	140
	Receiver ($/$m^2$)	91,000	57,143	65,563
	Power block ($/kW)	–	563	1040
	Thermal storage ($/$kW_{th}$)	88	49	22
	Total plant cost ($/kW)	–	6424	6506.9

[a]Current economic values
[b]Plant type: CSP power tower molten salt, PPA single owner, Station ID 91,486 (Daggett)

receiver and solar-to-electric efficiencies. The reduction of the solar collector cost per m^2 can also be highlighted.

For the case of solar towers, Table 3.6 shows data on different plants, whose comparison allows to approximate the advances taking place in the sector (own elaboration from SL [100: 5-1/5-2, 5-9/5-10, 5-32/5-33 and 5-40]) and SAM [98]. Despite the fact that the considered plants are very different, there are some important results as, for example, the increase in heliostat size and receiver efficiency.

Over the recent years, the activity of the sector has focused on addressing some existing challenges of CSP technologies. According to various authors, the following challenges are among the most relevant ones (see also Pitz-Pall et al. [1, 15: 86–128, 25: 10–50/10–52, 79: 31–34, 88, 94]):

- How to get resistant mirrors with high reflectivity (above 92% for wave lengths between 300 and 2500 nm) of low cost and easy maintenance (tolerance to periodic washing and low coefficient for the deposition of dust). This requires assessing different materials and simplifying the design of their supporting structures in order to reduce their weight, increase their rigidity, and simplify their assembling.

The reliability of the mechanisms for solar tracking has been improved and their costs have been reduced.

- How to reduce the losses in fluid temperature, whether due to the displacement of the fluid through the multiple conduits of the plant, failures in the sealing of joints and valves, etc.
- Mitigate the effect of intermittency (due to the cycle day/night) of the primary source of energy (DNI), as well as its irregularity due to the presence of clouds. This has led to the improvement of the hybrid systems and the TES. In this second case, the aim has been to reduce cost and minimize heat losses and its energy requirements.
- Increasing the working temperature of the plants, which lead to a greater efficiency. The efficiency of the turbines increases proportionally to the increase in the temperature of the steam that moves them. For temperatures up to 400 °C, only a few improvements in the performance could be expected, although the costs can be reduced. On the contrary, significant improvements have been observed as a result of high temperatures. *Ceteris paribus* a higher working temperature entails a lower size of the solar field. In turn, the thermal storage is also more efficient, since each unit of the fluid can generate more energy.

The systems' improvements have been gradual, and their effectiveness has had to be clearly demonstrated, before being commercialized.

The expectations for the next years are shown in Fig. 3.9. Expressed in \$2015, it is expected that the cost per kW for a parabolic trough plant of 160 MW and 7.5 h storage is reduced by a third by 2025, from \$5550/kW to \$3700/kW. The costs of a solar tower of 150 MW and 9 h of storage would be reduced by 37% by 2025 (from \$5700/kW to \$3600/kW) [42: 95, 96]. As it can be observed, the most relevant factors driving this reduction are the solar field and engineering, procurement, and construction (EPC) costs.

Regarding the LCOE, the calculations for 2015 are in the range between \$0.14/kWh for a DNI of 2900 kWh/m^2/year and \$0.19/kWh for a DNI of 2000 kWh/m^2/year [42: 90]. A LCOE of \$0.06/kWh for a parabolic trough plant and \$0.07/kWh for a solar tower (\$2015) are expected for 2025 [42: 99]. A similar expectation can be found in IRENA [44: 86, 45]: a weighted average LCOE of $\$_{2016}$0.072/kWh is forecasted for 2022. Those values are lower than the previously extrapolated ones. However, these sources strongly advise to consider and interpret these results with caution because LCOE and auction prices are different and these prices apply to projects that will be commissioned in the period 2020–2022 and beyond (see also [64]).

3.2.4 A Cost Comparison with Other Renewable Technologies

To complete the analysis developed in the previous sections, this one presents two figures and a table with the most recent LCOE estimations for the most important

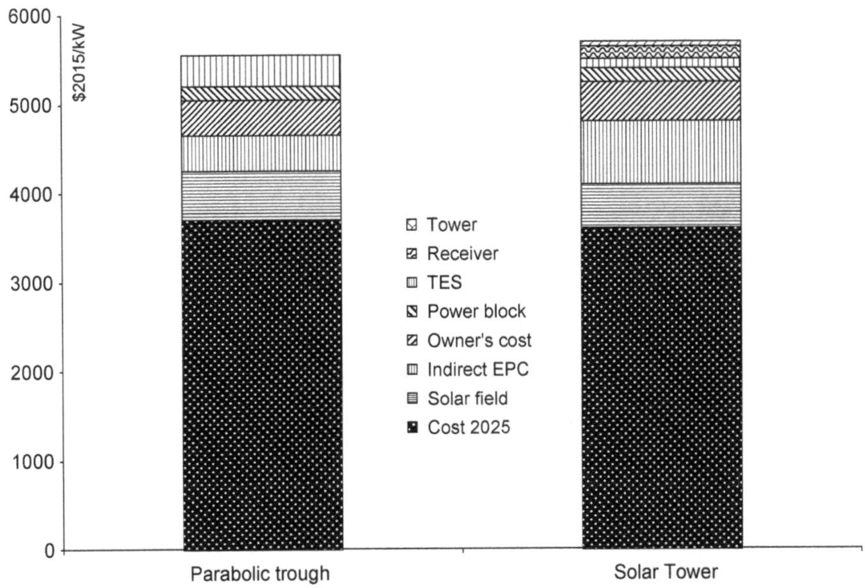

Fig. 3.9 Cost reduction in CSP plants 2015–2025 ($2015). *Source* Own elaboration from IRENA [42: 96]

Table 3.7 LCOE ($\$_{2016}$/kWh) renewable utility-scale plants

LCOE ($2016)	Biomass	Geothermal	Hydro	PV	CSP	Offshore wind	Onshore wind
2010	0.07	0.05	0.04	0.36	0.35	0.17	0.08
2017	0.07	0.07	0.05	0.1	0.22	0.14	0.06

Source IRENA [45]

renewable energy technologies [44–46]. The global weighted average LCOE (in $2016) of onshore and offshore wind generation, PV and CSP, is shown in Fig. 3.10. Observed data cover up to 2017 while, for the years 2018–2021, data come from extrapolations made by the authors of the cited works (except the values for onshore wind in 2021, off-shore wind in 2018, PV in 2020 and 2021, and CSP in 2018 and 2021, which have been obtained by logarithmic interpolation). As it can be observed, solar technologies show a rapid reduction of the LCOE, while wind technologies show a slower one. However, all of them converge in a narrow band ranging from $0.0457 kWh to $0.08/kWh.

Table 3.7 contains the weighted average values of the observed LCOE ($\$_{2016}$/kWh) related to renewable utility-scale plants.

Despite the reduction in the costs of CSP generation, this was the technology with the highest LCOE in 2017. Therefore, it is expected that the CSP LCOE will further decrease in the next few years. This is shown in Fig. 3.11. The figure shows

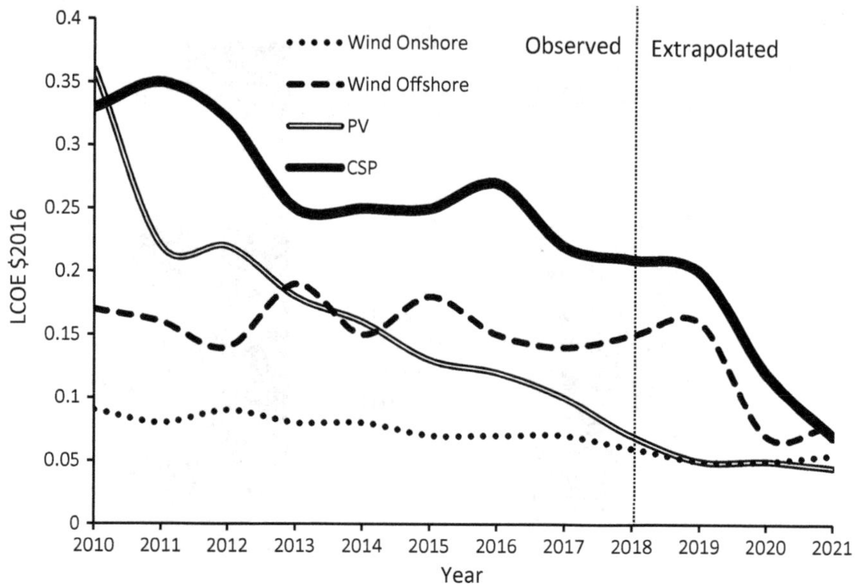

Fig. 3.10 Cost reduction of renewable energy plants ($2016). *Source* Own elaboration from IRENA [45]

the indexes (2010 = 100) of the weighted average LCOE reductions of the main renewable technologies throughout the present decade. As noted, solar technologies show the sharpest cost decline, although this reduction is more regular in PV and has ups and downs in CSP. In this case, there are two stages: 2010–2016, with a ~20% LCOE reduction and 2016–2020, in which a ~70% drop is estimated. Nonetheless, the forecasts contained in Figs. 3.10 and 3.11 depend on many factors. There is no doubt that for CSP generation, in addition to the technical performance improvements, the DNI level of the plant site and the project financial conditions play a crucial role. In recent years, geographical knowledge for the best location of CSP plants has greatly improved.

3.3 Drivers and Barriers to Concentrating Solar Power Deployment

The development and deployment of new, low-carbon technologies are an essential part of efforts to mitigate climate change. However, historical trends are clear: Energy technologies do not emerge and diffuse quickly due to a wide array of barriers to invention, development, and diffusion [29, 33]. Policymakers need to identify ways in which the process can be accelerated [68]. Indeed, experiences in different countries show that diffusion can be a very slow and tedious process [77], particularly in the

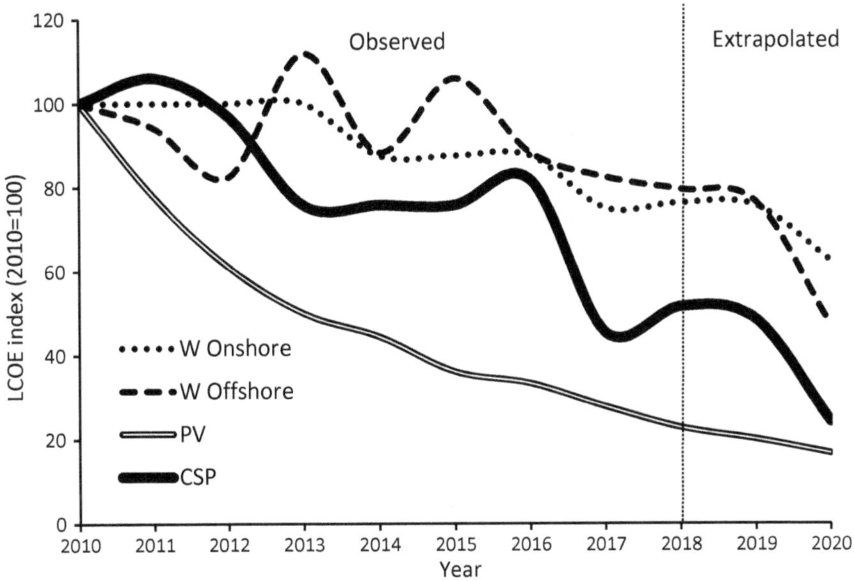

Fig. 3.11 Index of cost reductions of renewable energy plants (2010 = 100). *Source* Own elaboration from IRENA [45] data, except 2018 and 2019 values which have been obtained by logarithmic interpolation

energy sector. As Rosenberg [95] puts it, at the beginning the new technology is crude, expensive, inefficient, and badly adapted to the existing institutional setting and the ultimate use, which leads to slow diffusion.

3.3.1 A Literature Review on the Drivers and Barriers to CSP Deployment

Identifying the drivers and barriers to renewable energy technologies in general and CSP deployment in particular is a relevant exercise in order to propose policy measures which activate those drivers or remove those barriers. Compared to intermittent renewable energy technologies, CSP has a main distinguishing feature: It can be equipped with low-cost thermal energy storage, which allows it to provide dispatchable renewable power. It can then be a cost-effective, flexible option in different places, especially with increasing shares of variable renewable electricity [46, 70].

This section reviews past studies on the drivers and barriers to CSP deployment. While many studies have analyzed the drivers or barriers to the diffusion of CSP, many of them have focused on a particular case without taking a comprehensive view of the topic.

The review was carried out by looking into journal articles, official statistics, reports from industry associations, research organizations and other institutions (the European Commission, IRENA, Protermosolar, ESTELA, and IEA, among others), and news items from newspapers, government and company Web sites were reviewed. The most relevant energy journals were consulted (including the electricity journal, energy policy, energy journal, energy, renewable and sustainable energy reviews, energy economics, energy journal, solar energy, applied energy, and nature climate change and environmental economics and energy policy). In addition, publications exclusively dedicated to CSP (CSP Today and Helio CSP) were consulted. Furthermore, a general Google search for documents in the gray literature was undertaken. Our review covers the last ten years (2008–2017). This review partly draws on the work and findings of del Río and Kiefer [21], who focus on the EU. In contrast, we expand the geographical scope of the review and take into account the drivers and barriers to CSP deployment all over the world. It should be taken into account that the studies included below refer to some barriers in the past which were context (country)-specific and that do not necessarily reflect the current siuation in the country with respect to CSP or the present status of this technology (including its maturity and costs).

Islam et al. [47] provide a holistic review of CSP technologies by (1) analyzing the present global status of CSP technology implementation, (2) identifying major research findings of previous review articles, (3) discovering the historical development and recent trends on CSP research. They suggest that competition with PV has been harmful for CSP in the past. According to the authors "due to low cost of solar PV, many of the investors in CSP technology were moving towards the technology, however, there is a potential for integrating a solar tower with solar PV" [47: 995]. They also emphasize the importance of having good resources (land, water, and DNI) for CSP deployment. Capital cost, operation and maintenance cost, efficiencies, land and water requirements are compared. For all technologies, solar radiation, land and water requirement was found to be 1800 kW h/m^2, 5–7 acres/MW and 4 m^3/MWh, respectively [47: 999].

Maturity and efficiency of plants are also mentioned by the authors. Parabolic trough technology is the most proven, solar tower is mature, and both paraboloidal dishes and linear Fresnel reflectors are in the demonstration stage. Solar towers and paraboloidal dishes are regarded as the most efficient CSP plants, expected to have a 50% better efficiency than the trough and the Fresnel plants. Even though paraboloidal dishes have the highest efficiency, the basic plant cost, the operation and maintenance cost, and the capital costs are the highest among the plants, with the linear Fresnel reflectors being the lowest [47: 999].

The authors also focus on a particular CSP design (solar tower), arguing that, in order to reduce the financial risk and to lower the cost of electricity production, solar tower plants (i.e., commercial plants with a capacity of >30 MW) should often hybridize with natural gas combined-cycle, coal-fired, or oil-fired Rankine plants.

Unlike other CSP plants, solar tower plants require a considerable water supply and the largest land area, and they must be large in order to be economically viable and profitable (between 50 and 100 MW) [47: 905].[27]

Bijarniya et al. [7] provide a review of CSP technology in India. The authors identify several barriers to this technology in this country, including lack of technological data, financial barriers, as well as complex administrative and bureaucratic issues. They stress the inhibiting role of PV as the competitor of CSP due to dramatic reductions in cost, lack of government support and policy initiative slowing down CSP progress. However, they also identify a driver: CSP plant can be hybridized with the existing coal-based power plants.

Gauché et al. [31] focus on the system value of CSP. They consider several key items contributing to the value proposition of CSP (see Table 3.8). The authors also note the existence of several barriers which are beyond the technology but which CSP needs to overcome in order to be deployed [31:137], without ranking their importance:

- Increased cumulative operating experience at plant level for CSP with storage.
- Advance along the learning curve, through both a sustained growth rate and knowledge sharing between the developers.
- Improving the technology to make bankability easier. Increasing the operating experience as well as the modularity in order to reduce the amount of up-front costs is deemed particularly relevant in this regard.
- Proving systems and integration value through demonstration and a refined systems analysis.
- Enhancing social and environmental acceptance by factoring in all complexities and societal feedback. In particular, the authors pay attention to social and environmental issues which might be an obstacle for the diffusion of the technology.
- Convincing policymakers to recognize the energy security value of CSP.

Haas et al. [36] analyze the barriers for the massive deployment of solar technologies in Chile, including both PV and CSP, based on data from 50 interviews with experts from industry, technology providers, academia, solar plant operators, and government. The authors identify several types of barriers: economic and financial (insufficient financing schemes and volatile energy prices), market (immature solar market and insufficient local products and market concentration), system integration (limited transmission capacities, backup flexibility and distant energy supply and energy demand centers), solar-technical (solar mapping and forecasting, harsh environment in terms of soiling, corrosion and degradation and access to water), regulatory (delays in the environmental assessment process, difficulties in getting land concessions and difficulties in grid connections) and information barriers (lack of technical skills and training institutes and lack of social awareness and social involvement). The main conclusion is that all those barriers are very country-specific.

[27]The efficiency of the plant varies, depending on a number of variables such as the optical characteristics of the heliostats, the accuracy of the mirror's tracking system, and the cleanliness of the mirror.

Table 3.8 List of key items contributing to the value proposition of CSP

Value proposition item	Description
Renewable and sustainable	– Low-carbon footprint over life cycle of the technology including low-to-zero carbon emissions during operation and – Offers system resilience to fuel price fluctuations
Thermal energy storage	Highest efficiency, large-scale storage availability enables high capacity factor and/or capacity credit
Ramp rate, turndown limits and dispatchability	– In combination with storage, fast ramping enables good grid operator control and the ability to serve electricity in peak times, particularly the evening peak – Heat transfer by HTF or storage medium permits the faster ramp rate and turndown limit, usually constrained by combustion process
Inertial electricity	The use of fairly traditional rotating heat engines offers voltage stability
Hybrid and multi-use options	The conversion from sunlight to centralized thermal energy allows for various combinations of energy inputs and/or energy uses to suit specific needs

Source Gauche et al. [31: 124]

Three interesting barriers are the lack of local players (a few local companies leading to an immature market and higher costs than in other countries), the lack of technical skills (lack of human capital along the whole value chain as well as training institutes), and the lack of social awareness and social involvement. The authors also propose “direct” and “indirect” measures to mitigate each of those barriers.

Schinko and Komendantova [97] analyze the risks of investments in CSP in North Africa and propose measures to de-risk those investments. They use an LCOE model and find out that comprehensively de-risking CSP investments lead to a 39% reduction in the mean LCOE from CSP. However, this reduction is still not sufficient to achieve economic competitiveness of CSP with highly subsidized conventional electricity from fossil fuels in North Africa. Hence, their results suggest that de-risking represents an important strategy to foster the deployment of CSP in North Africa, but additional measures to support renewable energy sources will be needed. The authors assess the impact of nine different risk categories (regulatory, political, revenue, technical, financial, force majeure, construction, operating, and environmental) on the LCOE of CSP. They show that, by de-risking investments into CSP projects in the North African region, lower financing costs and capital costs result, leading to a lower LCOE of CSP projects. This is so because the cost of capital is, by far, the most influential component of the overall LCOE [97: 267]. However, the authors warn that "not all components of investment risks for RES projects in the North African region will be avoidable in reality" [97: 266].

Kost et al. [56] analyze CSP plant investment and operation decisions under different price and support mechanisms. The authors assess the economic value of CSP storage via energy modeling of a Spanish plant location under the respective wholesale market prices as well as the local feed-in tariff. The analysis shows that investment incentives for CSP plants with storage need to appropriately account for the interdependency between the price incentives and the plant operating strategy. Interestingly, the authors note that the current Spanish support scheme only offers limited incentives for larger thermal storage capacity [56: 238]. The authors show that, compared to feed-in tariffs, feed-in premiums allow CSP generators to better capture the value of their technology. They argue that investors will choose different plant layouts depending on the remuneration scheme and that a remuneration scheme which distinguishes between dispatchable and non-dispatchable renewable energy technologies should be implemented in order to encourage a greater amount of electricity generation from renewable energy sources [56: 247].

Frisari and Stadelmann [30] analyze the influence of policies and international finance institutions on the de-risking of CSP in emerging markets. The authors argue that, so far, costs were a major deterrent for CSP investments. The high cost of CSP is the main barrier to rapid deployment. The difference between the cost of generating power from CSP plants (around 0.2–0.3 USD/kWh) and the revenues that project developers can make in the electricity market was substantial: 98% of investments in CSP have needed some form of public support in both developed and developing countries [30: 13]. The high cost and perceived risks of CSP require public finance to improve the projects' financial profile [30: 21]. In addition, they highlight the importance of other barriers. Besides technology costs, investors face considerable technology, regulatory and financing barriers, particularly in emerging and developing economies. The limited experience with CSP in many of these countries increases technology risks, including the risk of solar resources being lower than predicted. CSP projects face further financing risks in North African countries as their financial markets are often not fully developed or well suited for project financing, offering high-interest rates and short maturities on debt [30: 13]. They also argue that the higher value of CSP compared to other technologies could be a driver for its diffusion.

San Miguel and Corona [96] evaluate the economic viability of a representative CSP plant deployed in Spain (50 MW parabolic trough plant with 7.5 h of thermal energy storage).[28] Their results show the limited competitiveness of this configuration, which is attributable to its high capital costs, high fixed operating costs per unit of output and the limited revenues from power sales. The authors argue that, although the intrinsic limitations of the technology are also to blame (such as the scale factor and limited modularity of thermal power plants, or the limited availability

[28]The subsidies applied during the second half of the 2000s caused a rapid expansion of the CSP sector in Spain with the construction of 50 commercial plants. Most of this growth (96.5% of installed capacity) was based on replicating that specific configuration.

of DNI resources in locations with high electricity demand), the main reason for the slow development of CSP is mainly related to its high generation costs [96: 205].[29] They find out that, with exceptions, the policy strategy followed in Spain had limited success at promoting technology advances with the potential to achieve higher generation capacity, improved revenues, reduced costs and increased dispatchability [96: 205]. The authors argue that, although dispatchability can be a driver, the limited dispatchability of the aforementioned specific CSP configuration has actually been a barrier.

Under the financial and technical conditions considered by the authors, this type of CSP configuration would require approximately a fourfold increase in revenues to make the project economically attractive. This may be reached through direct public subsidies (a premium on top of market price) or as a result of higher average electricity prices. This is the result of the limited dispatchability of this CSP configuration and the reduced improvement margins that are attainable. According to the authors, CSP may be better suited in remote or isolated locations inaccessible to conventional grids, where electricity prices are not governed by conventional market mechanisms and may be significantly higher [96: 207].

Bijarniya et al. [7] perform a Strengths, Weaknesses, Opportunities, and Threats (SWOT) analysis of CSP in India, focusing on the analysis of the barriers to CSP deployment. The authors argue that, although current CSP technologies are capital-intensive and have a high LCOE compared to other available renewable-based technologies like PV, wind energy, and biofuels, they have many advantages with respect to them [7: 594]. These include (1) a higher reliability of CSP electricity due to its storage capability; (2) ease of integration of CSP technologies in present infrastructures with similar operational characteristics, like coal/gas-based thermal power plants; (3) CSP power plants can be set up easily, since their components are made up of common industrial materials such as steel and glass. The authors found that several barriers like insufficient DNI data, improper policies, complex land acquisition systems, availability of water, permitting issues, and expensive finance are responsible for slow growth of the CSP technologies. Financial barriers and the reduction in the costs of PV modules are also major obstacles in promoting CSP. Government policies, proper maintenance, and lack of skills are the main obstacles to CSP development in India [7: 601].

Naik et al. [76] identify several barriers to solar thermal technologies in India (both for heat and power generation) and classify them in several categories. They distinguish between technology (immature and inefficient technology, unreliable, uncertain and sometimes incompatible technology and unavailability of skilled manpower), economic (high investment costs, relatively high payback periods, difficulty in obtaining bank loans and lower government support), institutional (lack of infrastructure, lack of purchasing power, lack of coordination among different agencies

[29]The authors stress the main differences with respect to PV: In contrast to PV cells, which may use all forms of solar radiation (direct, diffuse, and reflected), the primary energy resource for CSP is DNI. This type of radiation prevails in subtropical latitudes, usually coincidental with desert climate zones. Another key difference with PV relates to the limited modularity of CSP. Due to the thermal nature of the technology, CSP plants need large scales to achieve high efficiencies [7: 206].

and poor technological transfer) and social/cultural barriers (unwillingness in adjusting to change, lack of awareness and motivation, disharmony with social or cultural belief, and lack of perceived needs).

Peters et al. [86] perform a qualitative assessment of different solar technologies, focusing on the comparison between PV and CSP parabolic trough. Several parameters and variables are considered: (1) technological uncertainty; (2) cost transmission over 3000 km; (3) storage potential; (4) resource bottlenecks in terms of water and material for key components; (5) addressable market (modularity, geographies, combination with fossil-based power plants, slope angle restrictions, and side products); (6) environmental impacts (lifecycle greenhouse gas emissions, toxicity, and land use), and (7) local value creation/employment opportunities. Their qualitative assessment does not lead to a clear competitive advantage for any technology. Thus, the authors suggest that a variety of technologies should be maintained and developed in order to mitigate the risk of picking the "wrong" design.

Purohit and Purohit [91] analyze the barriers to CSP generation in India. The authors find out that the availability of long-term solar radiation data across the country is one of the most important technical barriers for the financial closure of the solar power projects. Additionally, limited meteorological information, land availability, water availability as well as grid loading and availability were bottlenecks experienced by CSP projects in this country [91: 650]. The authors argue that competition with PV, and particularly its costs, has been an important barrier for CSP,[30] but there have been other obstacles, including expensive financing, unclear future of government subsidies, "need for a local manufacturing, and the tight profit margins and even stiffer time limitations" [91: 662]. For these authors, "the primary barrier to utility-scale solar power is project financing. High up-front cost of CSP projects increases LCOE as compared to fossil fuels. This is a significant barrier as financiers are unfamiliar with CSP investments, risk-averse, and often focus on the short term" [91: 662]. However, the authors identify several drivers for future CSP deployment in India: the increased availability of measured DNI data, the increasing local manufacturing capabilities, and policy changes [91: 663].

Labordena et al. [57] have analyzed the impact of political and economic barriers on dispatchable electricity from CSP plants equipped with thermal storage in sub-Saharan Africa. They show that electricity from CSP is generally not competitive with coal power in this region, even considering expected cost reductions up to 2025.[31] They stress the role of investment risks, which arise because of political, regulatory, financial and administrative barriers and long and uncertain permission processes [57: 54]. They also show that variations in risk across countries influence the cost of power from CSP more than variations in solar resources and claim that de-risking policies for CSP investment and policies to increase institutional capacity

[30]"As long as energy price of PV plants is less than the energy price of equivalent CSP, and continue to decline, PV will remain a preferable solution over CSP for energy investors" [91: 662].

[31]This is the case except in Southern Africa, where solar resources are excellent and financing costs are comparatively low [57: 60].

and cooperation among sub-Saharan Africa countries could reduce costs and make CSP cheaper than coal power by 2025 in all sub-Saharan Africa countries.

Xu et al. [120] analyze the drivers and barriers to CSP generation in desert regions. The authors identify key drivers in this regard and, particularly, the potential of CSP for power generation throughout the day for base load applications and varied applications. However, they also show that there are critical barriers: the enormous amounts of support and capital investments being required and challenges related to water consumption, materials design and development for the optimum heat transfer fluid, thermal energy storage and receiver subsystems in addition to commercial viability and environmental impacts [120: 1106].

Lilliestam and Pitz-Paal [64] identify possible drivers of CSP deployment in their analysis of the cheap PPAs for two recent CSP projects in Australia (Aurora) and Dubai (DEWA IV). They investigate reasons for the low PPAs and find that both projects have low technology costs. However, the authors argue that this is an insufficient explanation for the low bids. Therefore, they look for additional factors. For Aurora, a key explanation is its business model that allows it to sell power outside the PPA, during high-price times when the sun sets and the growing PV fleet does not generate electricity. For DEWA IV, the two key factors are its long PPA duration and very low financing costs. They conclude that both projects can probably be replicated, either in places with an increasing PV fleet and strong "duck curve" problems (Aurora) or in places with low policy risks and access to cheap capital (DEWA IV). Such places could include the USA, Southern Europe, or the Gulf region [64: 17].

Mahia et al. [66] conduct an in-depth survey with senior managers of CSP companies in order to identify which specific components of the production process and technology could be manufactured in Morocco, and which changes would be required for production to be realized. In addition, they studied the barriers that currently hamper the development of the renewable production industry and the relative advantages offered by Morocco and the MENA region for CSP sector development. They categorized these barriers into three major groups: entrepreneurial, policy- and market-related and asked all the experts to rate each barrier using a scale of relative importance in Morocco. They find out that policy-related barriers are more relevant than entrepreneurial or market barriers.

Regarding the drivers (or so-called opportunities/advantages), the following are the most relevant (in decreasing order of importance): High solar potential (irradiation), political/institutional will to increase RES (CSP) potential, high level of European commitment for the development of RES potential in the region, low labor costs, high potential growth in electricity demand, political/institutional will, at the country level, to promote a local RES technological industry, strong economic growth perspectives, high level of multilateral commitment for the development of renewable energy sources potential in the region, export potential of manufactured renewable generation plant components from the country to the rest of the region, successful previous initiatives (CSP experience and know-how in MENA) and export potential of manufactured renewable generation plant components to the rest of the world [66: 594].

Dowling et al. [24] review economic studies of CSP systems from the perspective of investors and operators. The authors find that CSP technologies may remain unattractive from a LCOE perspective relative to other renewable technologies (e.g., wind, photovoltaic). A key advantage of these technologies, however, is the relative ease of energy storage and dispatchability, which is not captured by the LCOE calculations. The authors conclude that it is thus critical to analyze CSP systems with alternate economic metrics [24: 1025]

Köberle et al. [52] assess the current and future techno-economic potential of CSP worldwide, comparing it with PV electricity generation. The authors find that CSP is more competitive in desert sites with highest direct solar radiation. PV is a clear winner in humid tropical regions and in the temperate northern hemisphere [52: 739]. The four main conclusions of the study are: (1) there is a substantial potential for both CSP and PV; (2) CSP and PV compete for the same resource (land); (3) the remoteness of many sites with some of the best potential raises the question of transmission costs to demand centers[32]; (4) CSP electricity generation costs might in the medium term decline faster than PV. An alternative scenario explored assumptions consistent with considerable support in all phases of RD&D capacity deployment to speed-up technology development [52: 753].

Lilliestam et al. [60] compare and contrast carbon capturing and storage applied to coal-fired power plants with CSP. The authors stress that CSP can provide base load electricity but also that it is in the early stages of maturity and it is more expensive than existing electricity-generating options, although costs should decrease with large-scale deployment. Barriers to scaling up quickly include it is more expensive than conventional power; the initial support needed; that investments in generation and transmission are required; the uncertain transmission system and that an EU treaty with North Africa and international power market is needed [60: 452]. The need for international cooperation may impede CSP expansion in Europe. In contrast, water is unlikely to severely limit overall expansion.

Komendantova et al. [53] analyze the risk perception of CSP projects in North African countries. Twenty-three experts from industry, government, the financial sector, and the scientific community were interviewed. More than half of all respondents identified the complexity of bureaucratic procedures and corruption as significant barriers. Accountable and capable bureaucracies help reduce transaction costs for entry, operation, and exit. Transparency could reduce uncertainties by providing more predictable application of government rules and regulations. Other risks identified as significant barriers were the instability of national regulations, the absence of guarantees from national governments, and the international community on invested capital and revenues from projects, a low level of political stability, and the lack of support from local governments as a result of a low level of awareness about the advantages of renewable energy sources [53: 106]. Stakeholders were concerned about three classes of risks, namely regulatory, political, and force majeure. Among them, regulatory risks caused the biggest concern by far [53: 107].

[32]Including estimates of the cost to build new transmission lines raised the cost of electricity generation in many of the very best sites worldwide.

Komendantova et al. [54] focus on the risks faced by developers of CSP projects in North Africa and the Middle East. The authors find that project developers are most concerned about the risks related to corruption as well as inefficient and unpredictable bureaucracies, rather than with terrorism and rogue state behavior. Those risks lead to cost overruns and delays in the planning and construction phases of project development, negatively affecting its profitability [54: 4835]. Several policies can reduce such risks, either by mitigating their consequences or by seeking to eliminate their underlying causes. The authors suggest measures such as supply of insurance by international organizations and public–private partnerships. In the long term, the authors argue that it will be important for North African countries to take steps to improve their own government accountability. The authors argue that addressing the aforementioned risks will take concerted action by policymakers from both North Africa and Europe.

Medina et al. [69] have applied a Scenario Analysis to identify the barriers affecting the decision to invest in the CSP sector in Morocco. These barriers are the ones perceived as more important by international CSP companies regarding a future 10-year scenario in Morocco. The results show that the companies without a presence in Morocco perceive uncertainty, insecurity, and informality as the most important barriers. Uncertainty and insecurity are related to the need to achieve a critical level of market development to start a business in the sector, the higher capital costs (risk premium) for initiatives in the area, the social and political instability in the region, and the insufficient long-term security for planning [69: 50]. In contrast, financial, legal, and market risk are the main concerns among those companies that are active in Morocco [69: 50].

Bosetti et al. [8] use an expert elicitation to identify the future costs of PV and CSP as well as the barriers to the uptake of these technologies. Sixteen leading European experts from the academic world, the private sector, and international institutions were interviewed. One key insight is the importance of appropriately supporting the full RD&D process. The experts indicated that demonstration activities should be a core element in the innovation strategy for solar technologies [8: 316] and that non-technical issues and obstacles could slow down the worldwide diffusion of solar technologies [8: 314].

Kaygusuz [51] assesses the potential of CSP in Turkey and provides strategies to promote the development of this technology. The author argues that technology and cost are two major barriers to CSP development in Turkey. In contrast, the solar resource is abundant and land availability will not become a barrier in the future. The solar resources and large wasteland areas are widely available in the western and southeastern part of the country. The relevance of public support (for deployment and RD&D) is stressed.

Trieb et al. [107] aim to provide a strategy for the market introduction of CSP plants in the MENA region without considerable subsidization. The authors argue that there is a business case for CSP in this region. Their model analysis shows that even a cost of CSP of around ¢28/kWh can be a least-cost option when compared to the average cost of power production of around ¢8/kWh, which usually includes peak power at a significantly higher cost [107: 316]. They emphasize the importance of credibility, the ability to plan future capacity additions, and the stability and reliability of the legal, political, and economic framework conditions in order to reduce investment risks and to trigger local manufacturing industries [107: 312]. CSP plants can replace peaking power, medium load power, and base load power if specifically designed for these functions [107: 316]. The authors propose long-term power purchase agreements using tariffs that fully cover the cost of CSP.

Pietzcker et al. [87] focus on two main aspects of drivers and barriers to CSP: dispatchability (as a driver) and the competition of CSP and PV (as a barrier). Using the integrated energy-economy-climate model REMIND, they show that PV is cheaper on a direct technology basis and is thus deployed earlier. However, the authors stress the relevance of integration costs for the competition between PV and CSP. "Although PV consistently has lower direct LCOE than CSP and initially deployed faster, CSP catches up and overtakes PV at the end of the century due to lower integration costs of CSP" [87: 718].

Many papers focus on dispatchability as a driver, including Pietzcker et al. [87] and Forrester [28], among others. The survey of Forrester [28: 1640] leads to the conclusions that there is a reasonable degree of convergence in the results of quantitative studies of the system costs and benefits of CSP with thermal energy storage and that utilities and regulators around the world are beginning to calculate net system costs when valuing alternative renewable resources.

Del Río and Kiefer [21] have provided an integrated analytical framework to identify the drivers and barriers to CSP deployment. The authors empirically identify those drivers and barriers to CSP deployment in the EU in the past and the future with the help of a literature review. Text, they rank those drivers and barriers according to the views of investors and other relevant stakeholders involved in CSP. Whereas their review of the literature suggests the relevance of a wide array of drivers and barriers, the empirical analysis performed, based on an expert elicitation and an investors' survey suggests that the degree of importance of each driver/barrier differs for different types of stakeholders (industry, researchers, policymakers, and others), different time frames (past and future) and different CSP designs (parabolic trough and solar tower). Regarding the past drivers of CSP deployment, the expert interviews have suggested the importance of deployment support, policy framework conditions, and policy ambition and the technology being regarded as proven (technology risks). Dispatchability is regarded as the main future driver of the technology, followed by policy framework conditions and policy ambition and complementarity with PV. The investors' survey confirms the relevance of dispatchability as a driver, together with key technology features (maturity and good performance of the technology) and investors' features (accumulated knowledge and experience) specifically for the case of parabolic trough. Regarding CSP deployment in the past, several barriers stand out. These include higher costs, retroactivity, lack of stability, and ambition of targets and low levels of deployment support. Higher costs, limited resource potentials (DNI) and retroactivity, lack of stability and ambition of targets are perceived as the most relevant future barriers for experts. The view of investors on those barriers is significantly different. They stress the importance of administrative processes, construction permits, and grid connection. In short, the views of investors and experts regarding both drivers and barriers are deemed complementary, since they focus on different levels of analysis.

The studies mentioned and reviewed above suggest the existence of different categories of drivers and barriers, which in turn include different factors. Next section lists and describes those factors (according to del Río and Kiefer [21]).

3.3.2 The Drivers[33]

3.3.2.1 Techno-economic

Proven Technology (Low Technology Risks)

Technology risks are inherent to complex technology systems. The more mature and proven a technology is, the more attractive it is for potential adopters, which do not have to face the additional risks and costs of early adopters.

Cost Reductions

Since one main barrier to the diffusion of CSP may have been its high costs (see [23]), cost reductions are obviously the main driver for this technology. Cost reductions are due to several factors, including economies of scale, learning effects at both the industrial and plant levels, increased size, and technological improvements due to innovation. The first two are the result of deployment, whereas innovation is both the result of RD&D and, to a lesser extent, deployment. Several contributions suggest that there have been and will be substantial cost reductions for CSP. IRENA [44] estimates that total installed costs of newly commissioned CSP projects have fallen by 27% in 2010–2017. 37 and 43% LCOE reductions are expected for parabolic trough and solar tower, respectively, in 2015–2025 [42]. Recent auction results for CSP projects that will be commissioned after 2020 show costs falling to between \$0.06 and \$0.1/kWh [42: 16].

Improvement of the Technology Over Time

The technology, with a long development journey, has already reached the commercial stage. However, it is only at the beginning of its commercial deployment in terms of installed capacity. Therefore, a high technological dynamism and significant improvements and cost reductions can be expected in the future. On the other hand, innovation theory predicts that at the early stage of a technology, different designs compete between each other. This might also be the case with CSP, which has different designs (parabolic trough, solar tower, Fresnel and Stirling), although some experts would disagree that they compete between each other and even that they should be presented in equal terms. Within the different CSP technologies, there are different maturity levels. One design has been dominant (trough), but solar towers are expected to capture an increasing share of the market in the future.

[33]This subsection and the next heavily draw on del Río and Kiefer [21] and del Río et al 23.

Dispatchability and Higher System Value of CSP
The benefits of the technology for the adopter make a technology attractive. CSP has a very attractive feature in this regard. CSP plants with thermal energy storage allow higher capacity factors, dispatchability, contribute to grid balancing, spinning reserve, and ancillary services. They also have the ability to shift generation to when the sun is not shining and/or the ability to maximize generation at peak demand times [118: 31]. For this reason, it has a higher system value compared to other intermittent renewable energy sources (see Chap. 4).

Development in Niches
Niches provide a space for technologies to improve their performance through learning by using and interacting and through economies of scale [23]. Co-generation for domestic and industrial heat use, water desalination, and enhanced oil recovery in mature and heavy oil fields are other possible applications of CSP plants which are additional to electricity generation [40]. Hybridization with other technologies can also be considered a niche market for CSP technologies.

Complementarity with PV
The value of CSP will increase further as PV is deployed in large amounts, and, thus, they may complement each other.

Existence of a Dominant Design (Parabolic Trough)
The existence of a dominant design creates security for their investors and reduces the perception of risks of the technology, since this looks more reliable and mature. Immature technologies often do not have a dominant design.

3.3.2.2 Policy/Political

Framework Conditions and Policy Ambition
Framework conditions refer to those aspects of RES-E support that are either outside the support system itself or that may be designed similarly irrespective of the type of system applied [6: 133, 20], including grid access procedures, permit procedures, the existence of long-term targets, or investment security.

Design Electricity Market/System
Some designs of the electricity system, in which the dispatchablity of electricity generation technologies is considered and valued, may be more favorable for CSP.

Deployment Support
Regarding support instruments, two main categories can be considered: RD&D policies (at EU and member state level) and deployment support (at member states level). Both may lead to technological improvements and cost reductions. Several well-known promotion schemes for renewable energy deployment exist, which could also be applied to support CSP, including feed-in tariffs and feed-in premiums, whether administratively—set or set through auctions, quotas with tradable green certificates soft loans, and investment subsidies (see [74] for a detailed description).

RD&D Support
Support for research, development and demonstration (RD&D) can be a driver of the technology since it leads to improvements and cost reductions. This support can be provided in several ways, e.g., support to industry (e.g., fiscal incentives), to public research centers (direct RD&D support), and to innovative demonstration plants within a public-private collaborative framework. Other policy interventions may favor networking and collaboration between private and public actors.

Regional Policies
Regions may provide support to CSP plants either directly (i.e., investment support) or indirectly (streamlining of administrative permits).

Carbon Prices
Carbon prices (whether from emissions trading schemes or carbon taxes) aim to internalize the negative environmental externality related to GHG emissions. Compared to conventional electricity generation, renewables in general and CSP in particular do not emit GHG during its electricity generation phase. Therefore, with a carbon price an extra cost is faced by the former, which makes renewables more competitive. Whether this is so depends on the level of carbon prices which, with the EU emission trading scheme, has remained very low.

Cooperation Mechanisms of the RES Directive
The cooperation mechanisms of the RES Directive may encourage the deployment of CSP. Cooperation mechanisms do not only bring greater flexibility for member states with low potential and/or expensive generation costs to partially meet their national targets in other countries, but also reduce the overall costs to realize the 20% EU RES target in 2020.

Planning Reliability (vs. Non-EU Countries)
Juridical security regarding administrative procedures in the EU may have been an attractive feature of investing in the EU versus investing in non-EU countries.

3.3.2.3 Social Acceptability

Social Acceptability
The social acceptability for a technology can be critical for its deployment (i.e., directly) but also to adopt policies which support it (i.e., indirectly). People might value that CSP technology deployment may provide substantial local value addition through localization of production of components, services, and operation and maintenance, thus creating local development and job opportunities.

3.3.2.4 Supply Chain Related

Local Manufacturing Capabilities
Thermal solar power plants demand regular industrial materials. Countries may pos-

sess a mature range of industries in the production of components and equipment for electro-thermal conversion so that an important part of the value chain can be added locally [115]. Having a well-developed local industry for components would make it easier to have access to those components for plant developers.

Strong Supply Chain
The presence of several capable actors in each stage of the value chain and the availability of standardized major components make the technology more attractive for potential investors.

3.3.2.5 Knowledge-Based

International Knowledge Collaboration, Information Flows
This refers to cooperation among research organizations in different countries and between those and industry. International knowledge collaboration leads to improvements of the technology, cost reductions and information flows, which may influence the speed of diffusion.

Strong Knowledge Base and Knowledge Generation in EU (vs. Non-EU)
Similarly, a strong knowledge generation base in the EU with respect to non-EU countries encourages the diffusion of the technology in the EU.

3.3.2.6 Resource Availability-Related

DNI Levels
Higher DNI levels obviously lead to lower generation costs for the same level of installed capacity. Therefore, places with higher DNI levels are more attractive for potential investors. This factor could be regarded as a precondition rather than as a driver.

Availability of Land
Availability of land in the South of Europe and, particularly, in Spain (with a low population density) may have been an important precondition for CSP deployment in the EU.

3.3.3 The Barriers

3.3.3.1 Techno-Economic

Technology Risks
Problems regarding performance of the technology would make it unattractive for potential investors and, thus, slow down its deployment.

Lower Technology Improvement than Expected
Unmet expectations about the improvement of the technology over time make it less attractive for potential adopters.

Existence/Absence of a Dominant Design
The fact that there are several technological alternatives may raise potential adopters' doubts about the virtues of the technology. According to this view, the absence of a dominant design is detrimental for the diffusion of the technology since it makes it less attractive for potential adopters.

Cost Comparison (Higher Costs)
Despite the aforementioned cost reductions in the past, the levelized electricity cost (LCOE) of CSP has been comparatively higher than for fossil fuel generation and other renewable energy technologies.

Lower than Expected and Uncertain Cost Reductions
Cost reductions may have been lower than initially expected. There was little change in the cost range for CSP projects between 2008 and 2012 (LCOE), although, since them, they have substantially been reduced (see above).

Competition with PV
Direct competition from PV is frequently mentioned as a potential barrier for CSP in the future [23]. Some authors argue that this competition may have delayed the deployment of CSP in some parts of the world.

Access to Credit
Access to credit to finance CSP investments may have been a barrier for the uptake of this technology in the past in the EU, and it may be so in the future. CSP is capital-intensive, financing costs represent a very relevant part of total costs and access to credit restrictions have occurred in the South of Europe (i.e., for any investment). According to Teske et al. [103: 93], "since the deployment of STE is still less than that of other technologies, private banks view these projects as higher risk, such that project financing has proven to be an obstacle for solar thermal electricity project developers in recent years. Project developers continue to have difficulties obtaining bank debt to fund their projects, due to the lack of long-term data on STE deployment and the irrational perception of STE as a risky and immature technology."

Impact of the Financial and Economic Crisis
The financial and economic crisis in the EU countries may have had a negative impact in the deployment of renewable energy technologies in general and CSP in particular. The economic crisis severely restricted the private sector capital that is used to finance RES-E projects.

Overcapacity and Meager Electricity Demand
One of the consequences of the economic and financial crisis in some EU countries has been a lower electricity demand than expected which, together with substantial investments in other electricity generation technologies in the early 2000s (e.g., Combined Cycle Gas Turbines) has led to overcapacities. Ceteris paribus this may

have been a barrier for the uptake of electricity generation technologies in general and CSP in particular.

More Attractive Investment Opportunities in CSP Outside the EU
CSP investment opportunities outside the EU have been increasingly attractive for a number of reasons (support from governments, policy mixes, good DNI). This leads investors to focus on those opportunities to the detriment of investments in the EU.

3.3.3.2 Policy/Political

General Legal Framework
These may negatively affect the uptake of the technology.

Design of Electricity Market
A design of the electricity system which does not value the dispatchablity of CSP would be unfavorable for this technology.

Retroactivity, Lack of Stability, Ambition of Target
These policy aspects can also be an important barrier for CSP deployment. Economic and political instability leads to higher risks and makes debt and equity financing more expensive.

Low Levels of Deployment Support
Low levels (or inexistence) of public support for deployment of CSP may have been a barrier to the deployment of CSP.

Low Levels of Support for Innovation and Demonstration
Low levels (or inexistence) of public RD&D support to CSP may be a barrier to the improvements, cost reductions, and knowledge accumulation required for the successful uptake of this technology in Europe (and elsewhere).

Difficulties in Using the Cooperation Mechanisms
Barriers to the use of cooperation mechanisms of the RES Directive would imply that CSP deployment would also not benefit from their use.

Administrative Procedures
Legal and administrative barriers (leading to long lead times for deployment and additional costs for project developers) are usually mentioned as a barrier to the deployment of renewable energy technologies in the EU. To our best knowledge, no study on the legal and administrative barriers specific to CSP is available, neither at the world nor EU level. According to IEA [39], difficulties in securing land, water, and connections and permitting issues have been barriers encountered by developers to establish CSP plants in some countries.

3.3.3.3 Social Acceptability

Local Opposition
Several local environmental impacts (land occupancy, leakages, water availability and impact on the landscape, particularly visual intrusion) may lead to a social backlash (not-in-my-back-yard) for this technology. There also risks for the personnel of the plant: the leakage of working fluid is potentially toxic. Poor knowledge about the technology (and its associated advantages over other RES) among different type of stakeholders, including policymakers (visibility gap) may be an indirect barrier.

3.3.3.4 Supply Chain Related

Weakness of Supply Chain
A narrow market problem in specific stages of the supply chain (few suppliers) may lead to a bottleneck in the supply for certain components and/or an excessive price for those.

Industrial Consolidation and Vertical Integration
Industrial consolidation (mergers and acquisitions) and vertical integration may lead to fewer actors in the supply chain and, thus, to a lack of competition in a specific stage of the process [23].

Unavailability of Standardized Major Components
Project-specific development may be necessary due to unavailability of standardized major components.

Exit of Large Players
Some large players have exited the market, whether for financial problems or other reasons. This could mean that the knowledge accumulated in those firms may also be lost, which would be detrimental for its further deployment. According to Lilliestam [63], this is a concern because several players have already left the market, leaving the current CSP market very thin, with only a handful of experienced firms active in each stage of the value chain. However, others believe that engineers from those firms have repositioned themselves in other companies.

3.3.3.5 Knowledge-Based

Low International Knowledge Collaboration
Few and non-intensive knowledge flows may be a barrier to the deployment of CSP (breadth and depth of cooperation).

Low Competence in the CSP Technology
Lack of skills throughout the supply chain and CSP technological innovation system may be a barrier to CSP.

Knowledge Generation Increasingly Moving Outside the EU
Knowledge about CSP has been accumulated in Europe, as a result of support for RD&D and deployment. However, the increasing deployment outside Europe, in addition to the stagnancy of CSP deployment in the European soil, may have also moved knowledge generation outside the EU, which could have a detrimental impact on CSP in Europe.

3.3.3.6 Resource Availability-Related

Limited Solar Resource Potentials
CSP plants can be only sited in areas with adequate solar resources, which restrict its potential deployment in Europe mostly to the Mediterranean area. DNI can reach 2000 kWh/(m^2a) in southern Spain which is high compared to other EU countries, but low compared, e.g. to the 2500 kWh/(m^2a) corresponding to the MENA region [56]. As a result, its highest growth potential is outside Europe, in the sunbelt region, which includes the Middle East, North Africa, South Africa, India, the southwest of the USA, Mexico, Peru, Chile, western China, Australia, southern Europe and Turkey [40].

Land Availability
CSP requires substantial space for its deployment. Land availability and competition for land use may have been and could be a hurdle in this context.

Water Availability and Competition for Water Use
CSP requires considerable water resources for its functioning. Water availability and competition for water use may have been and could be a barrier for the deployment of this technology in the past and the future.

3.3.3.7 Others

Risk of Environmental Pollution
Although, as a renewable energy technology, CSP is cleaner than its conventional counterparts, it still may lead to some environmental pollution (i.e., with oils). This concern could be a barrier for its deployment.

In addition, del Río and Kiefer [21] add some additional drivers and barriers specifically influencing investors such as resources, competencies and dynamic capabilities as well as previous experience accumulated in the firm which could influence whether companies invest or do not invest in CSP technologies (Table 3.9).

Table 3.9 Additional drivers and barriers at the investor level (resource availability and previous experience)

Driver/barrier	Explanation
Resource availability	
In their economic activity, firms are conditioned (constrained or enhanced) by their resource base, which comprises all resources, but also how these are put to use in daily business operations (competences) and how both are changed over time as a result of deliberate and dedicated action (dynamic capabilities). Together, these resources, competences, and capabilities form the firm's resource base, according to the resource-based view and its extensions	
Financial resources	The existence of adequate financial resources is a basic requisite for any investment. Financial resources may consist of available firm-internal funding or access to external funding and their corresponding conditions. Employing financial resources is attached to an expectation of return, both financially and strategically
Ownership of patents	Patents are the most tangible or observable form of knowledge. In sectors related to technology, knowledge is generally considered the most important firm resource
Availability of technological experience	Technological experience is the application of technological knowledge. It can initially be gained through demonstration of projects and later through regular ones
Skilled human resources	Knowledge and experience are deeply rooted in the personnel of firms. The learning organization is fundamentally based on learning human resources that interact and interchange knowledge. For the realization of complex technological projects, a skilled workforce is a prerequisite
Physical assets, such as installations and equipment	Physical assets determine the scale of business operations a firm can engage in. The existence and adequate use of special physical resources (laboratories, research facilities, or demonstration plants) related to experimentation and exploitation may give rise to new and innovative solutions
Engagement in collaboration networks	Cooperation is usually considered key in overcoming resource-based constraints in firms, as it may grant access to and use of resources, competences, and capabilities outside of the firm. This aspect has special relevance in highly complex technological projects, such as CSP plants. Collaboration networks may form around the value chain, around specific functions, such as knowledge creation, or center around specific activities. The deeper the collaboration is, the higher is a firm's institutional embeddedness in the corresponding (local, technological, etc.) clusters
Corporate image	Corporate image is determined by existing firm resources, competences, and capabilities and how these are perceived by third parties. Corporate image often acts as a proxy of the firm attractiveness, i.e., in purchase decisions (both by individuals and firms) or when initiating or during collaborations
Previous experience	
Firms can accumulate experience over time (i.e., it is "sticky"). On the one side, it increases the efficiency of business processes in firms, facilitating the realization of complex projects, on the other it may generate certain path-dependent trajectories that are exploited due to existing and increasing experience, creating situations of lock-in. Experience can cover wide ranges of domains	

(continued)

Table 3.9 (continued)

Driver/barrier	Explanation
Previous technology experience	As firms engage with specific technologies and technological configurations, they gain experience with these. Their use and application get more efficient. With time, selected technologies and configurations can become dominant, leaving alternatives behind
Previous market experience	Market experience covers aspects related to interactions along the value chain, both up-, down- and "side" stream
Previous project realization experience	Project realization experience comprises all bureaucratic and organizational steps from early project planning until functioning and includes, if applicable, decommissioning or management of the end-of-life of the project. Both internal and external aspects are covered
Previous investment in physical assets, such as other CSP plants or components	Despite being relatively easy to modify, change, or replace, the existence of physical infrastructure tends to produce lock-in effects (reluctance to change and self-reinforcing acting over the existing physical resource base, especially if newer infrastructure exists). Yet a certain level of physical assets is required when engaging in complex technological projects
Knowledge accumulated by previous CSP projects	Project experience can generate new knowledge or increase existing knowledge, including numerous aspects going beyond pure technological knowledge

Source del Río and Kiefer [21]

References

1. ATKearney (2010) Solar thermal electricity 2025. Clean electricity on demand: attractive STE cost stabilize energy production. A.T. Kearney Inc., Estela. Available at http://www.estelasolar.eu. Accessed May 2012
2. Baccani J (2014) Giovanni Francia. Nemo propheta in patria. Lo sfortunato destino della prima centrale solare del mondo. Article available at https://wsimag.com. Accessed May 2016
3. Baharoon DA et al (2015) Historical development of concentrating solar power technologies to generate clean electricity efficiently—a review. Renew Sustain Energy Rev 41:996–1027
4. Barbee J (2015) Could this be the world's most efficient solar electricity system? The Guardian, Wed 13 May 2015. Available at http://www.theguardian.com. Accessed May 2017
5. Behar O et al (2013) A review of studies on central receiver solar thermal power plants. Renew Sustain Energy Rev 23:12–39
6. Bergmann J et al (2008) Harmonization of support schemes. A European harmonised policy to promote RES electricity—sharing costs & benefits. A report compiled within the European research project futures-e (work package 3, Deliverable D17)
7. Bijarniya JP et al (2016) Concentrated solar power technology in India: a review. Renew Sustain Energy Rev 63:593–603
8. Bosetti V et al (2012) The future prospect of PV and CSP solar technologies: an expert elicitation survey. Energy Policy 49:308–317
9. Bradford T (2006) Solar revolution. The economic transformation of the global energy industry. The MIT Press, Cambridge
10. Braun FG et al (2011) Holding a candle to innovation in concentrating solar power technologies: a study drawing on patent data. Energy Policy 39:2441–2456

11. Burbridge D et al (n.d.) Stanwell solar thermal power project. Document available at http://citeseerx.ist.psu.edu. Accessed Oct 2016
12. Burgess et al (2011) Direct steam generation using the SG$ 500 m^2 paraboloidal dish concentrator. Research School of Engineering, Australian National University. Available at www.stg.anu.edu.au. Accessed May 2013
13. Butti K, Perlin J (1980) A golden thread—2500 years of solar architecture and technology. Cheshire Books, Palo Alto
14. CIEMAT (2015) Bringing Europe and third countries closer together through renewable energies. Better summary report. Intelligent Energy Europe (EU)/CIEMAT and others, Madrid
15. CSP (2007) Concentrating solar power. From research to implementation. Office for Official Publications of the European Communities, Luxemburg. Available at http://ec.europa.eu/energy/res/publications. Accessed Feb 2010
16. CSPGO09 (2009) Concentrating solar power: global outlook 09. Why renewable energy is hot. Greenpeace International with SolarPACES and ESTELA, Amsterdam. Available at https://www.greenpeace.org. Accessed Jan 2011
17. CSP Data (2017). Data on CSP stations, generation cost and many other issues available at www.csp.guru. Accessed June 2017
18. CSTP (2005) Concentrated solar thermal power-now!. Exploiting the heat from the sun to combat climate change. Greenpeace International with SolarPACES and ESTELA, Amsterdam. Available at https://www.greenpeace.org. Accessed July 2006
19. De la Tour A et al (2010) Economic analysis of the CSP industry. CERNA, Mines ParisTech and Agence Française de Développement, Paris. Available at http://www.cerna.mines-paristech.fr/Donnees/data07/704-CSP_report.pdf. Accessed Nov 2012
20. del Río P, Bleda M (2012) Comparing the innovation effects of support schemes for renewable electricity technologies. A functions of innovation approach. Energy Policy 50:272–282
21. del Río P, Kiefer C (2018) Analysis of the drivers and barriers to the market uptake of CSP in the EU. Deliverable 4.3 of the EU-funded MUSTEC project. http://mustec.eu/sites/default/files/reports/4.3Analysis_of_the_drivers_and_barriers.pdf. Accessed Dec 2018
22. del Río P, Mir-Artigues P (2012) Support for solar PV deployment in Spain. Some policy lessons. Renew Sustain Energy Rev 16:5557–5566
23. del Río P et al (2018) An overview of drivers and barriers to concentrated solar power in the European Union. Renew Sustain Energy Rev 81:1019–1029
24. Dowling AW et al (2017) Economic assessment of concentrated solar power technologies: a review. Renew Sustain Energy Rev 72:1019–1032
25. Drury E et al (2012) Chapter 10. Solar energy technologies. In: Hand MM et al (eds) Renewable electricity futures study. Volume 2: renewable electricity generation and storage technologies. Report NREL/TP-6A20-52409. National Renewable Energy Laboratory (NREL), Golden, CO, pp 10-1–10-60. Available at http://www.nrel.gov/analysis/re_futures. Accessed July 2014
26. Edenhofer O et al (eds) (2012) Renewable energy sources and climate change mitigation. Special report of the intergovernmental panel on climate change. Cambridge University Press, Cambridge
27. EurObserv'ER (2017) Solar thermal and concentrated solar power barometers. EurObserv'ER consortium. Release juin 2017. Available at www.eurobserv-er.org. Accessed July 2017
28. Forrester J (2014) The value of CSP with thermal energy storage in providing grid stability. Energy Proc 49:1632–1641
29. Fouquet R (2010) The slow search for solutions: Lessons from historical energy transitions by sector and service. Energy Policy 38:6586–6596
30. Frisari G, Stadelmann M (2015) De-risking concentrated solar power in emerging markets: The role of policies and international finance institutions. Energy Policy 82:12–22
31. Gauché P et al (2017) System value and progress of CSP. Solar Energy 152:106–139. Available at https://doi.org/10.1016/j.solener.2017.03.072. Accessed May 2018
32. Giménez J (2009) El régimen jurídico de las energías renovables en España. In: Becker F et al (eds) Tratado de regulación del sector eléctrico, vol 2. Thomson-Aranzadi, Navarra

33. Grübler A (2009) Transitions in energy use. In: Cleveland CJ (ed) Concise encyclopedia of history of energy. Elsevier, Amsterdam, pp 287–300
34. Günther M (2011) Advanced CSP teaching materials. Chapter 6: linear fresnel technology. EnerMENA and DLR. Available at http://www.energy-science.org. Accessed Apr 2017
35. Günther M et al (2011) Advanced CSP teaching materials. Chapter 5: parabolic trough technology. EnerMENA and DLR. Available at http://edge.rit.edu. Accessed Apr 2017
36. Haas J et al (2018) Sunset or sunrise? Understanding the barriers and options for the massive deployment of solar technologies in Chile. Energy Policy 112:399–414
37. Hernández-Moro J, Martínez-Duart JM (2012) CSP electricity cost evolution and grid parities based on IEA roadmaps. Energy Policy 41:184–192
38. IEA (2010) Technology roadmap. Concentrating solar power. OECD/IEA, Paris. Available at http://www.iea.org. Accessed Jan 2011
39. IEA (2014) Technology roadmap solar thermal electricity. International Energy Agency (IEA), Paris. https://doi.org/10.1007/SpringerReference_7300
40. IEA-IRENA (2013) Concentrating solar power technology brief, concentrating solar power technology brief. Available at https://doi.org/10.1063/1.2993731. Accessed June 2015
41. IRENA (2012) Concentrating solar power. Renewable energy technologies: cost analysis series. Volume 1: power sector, vol 275. Available at www.irena.org/publications. Accessed May 2014
42. IRENA (2016) The power to change: solar and wind cost reduction potential to 2025. International Renewable Energy Agency. Available at www.irena.org/publications. Accessed Mar 2017
43. IRENA (2017) REMAP renewable energy prospects for India. Working paper. Masdar Institute/IRENA, Abu Dhabi. Available at https://www.irena.org/-/media/Files/IRENA/Agency/Publication/2017/May/IRENA_REmap_India_paper_2017.pdf. Accessed Nov 2018
44. IRENA (2018a) Renewable power generation costs in 2017. International Renewable Energy Agency, Abu Dhabi. Available at http://www.irena.org/publications. Accessed Apr 2018
45. IRENA (2018b) IRENA renewable cost database and auctions database. International Renewable Energy Agency, Abu Dhabi. Available at http://www.irena.org/publications. Accessed Apr 2018
46. IRENA/IEA/Ren21 (2018) Renewable energy policies in a time of transition, Available at: http://www.iea.org/publications/freepublications/publication/IRENA_IEA_REN21_Policies_2018.pdf. Accessed Nov 2018
47. Islam MT et al (2018) A comprehensive review of state-of-the-art concentrating solar power (CSP) technologies: current status and research trends. Renew Sustain Energy Rev 91:987–1018
48. Jablonski S et al (2012) The Mediterranean solar plan: project proposals for renewable energy in the Mediterranean partner Countries region. Energy Policy 44:291–300
49. Jäger-Waldau A et al (2011) Renewable electricity in Europe. Renew Sustain Energy Rev 15:3703–3716
50. Jones G, Bouamane L (2012) «Power from Sunshine»: a business history of solar energy. Working paper 12-105. Harvard Business School, Boston. Available at https://hbswk.hbs.edu. Accessed Feb 2017
51. Kaygusuz K (2011) Prospect of concentrating solar power in Turkey: the sustainable future. Renew Sustain Energy Rev 15:808–814
52. Köberle AC et al (2015) Assessing current and future techno-economic potential of concentrated solar power and photovoltaic electricity generation. Energy 89:739–756
53. Komendantova N et al (2012) Perception of risks in renewable energy projects: the case of concentrated solar power in North Africa. Energy Policy 40:103–109
54. Komendantova N et al (2011) Solar power investment in North Africa: reducing perceived risks. Renew Sustain Energy Rev 15:4829–4835
55. Kost C et al (2012) Value generation of future CSP projects in North Africa. Energy Policy 46:88–99

56. Kost C et al (2013) Levelized cost of electricity renewable energy technologies study. Fraunhofer Institute for Solar Energy Systems ISE. Available at http://www.ise-fraunhofer.de. Accessed Oct 2015
57. Labordena M et al (2017) Impact of political and economical barriers for concentrating solar power in Sub-Saharan Africa. Energy Policy 102:52–72
58. Laffitte M, Massou F (2009) Le modèle de financement du Plan Solaire Méditerranéen. Réalités Industrielles, pp 27–35
59. Lilliestam J, Ellenbeck S (2011) Energy security and renewable electricity trade—will Desertec make Europe vulnerable to the "energy weapon"? Energy Policy 39:3380–3391
60. Lilliestam J et al (2012) Comparing carbon capture and storage (CCS) with concentrating solar power (CSP): potentials, costs, risks, and barriers. Energy Policy 47:447–455
61. Lilliestam J et al (2016) Understanding the absence of renewable electricity imports to the European Union. Int J Energy Sect Manage 10(3):291–311
62. Lilliestam J et al (2017) Empirically observed learning rates for concentrating solar power and their responses to regime change. Nat Energy 2:17094
63. Lilliestam J (2018) Whither CSP ? Taking stock of a decade of concentrating solar power expansion and development. Deliverable 4.2 of the MUSTEC project, European Commission under the H2020 programme, Zürich
64. Lilliestam J, Pitz-Paal R (2018) Concentrating solar power for less than USD 0.07 per kWh: finally the breakthrough? Renew Energy Focus 26:17–21
65. Lorec P, Schramm C (2009) Le Plan Solaire Méditerranéen: la dynamique d'un projet de coopération politique, énergetique et industrielle en Méditerranée et les défis à reveler. Réalités Industrielles, pp 15–19
66. Mahia R et al (2014) Assessing the future of a CSP industry in Morocco. Energy Policy 69:586–597
67. Martín H et al (2015) Promotion of concentrating solar thermal power (CSP) in Spain: performance analysis of the period 1998-2013. Renew Sustain Energy Rev 50:1052–1068
68. McDowall W et al (2013) The development of wind power in China, Europe and the USA: how have policies and innovation system activities co-evolved? Technol Anal Strateg Manag 25:163–185
69. Medina E et al (2015) Barriers to the investment in the concentrated solar power sector in Morocco: a foresight approach using the cross impact analysis for a large number of events. Futures 71:36–56
70. Mehos M et al (2015) An assessment of the net value of CSP systems integrated with thermal energy storage. Energy Proc 69:2060–2071
71. Meslier F, Palat P (2009) L'impact du Plan Solaire Méditerranéen sur l'évolution des interconnexions électriques entre les pays du pourtour méditerranéen. Réalités Industrielles, pp 20–26
72. Mills D (2004) Advances in solar thermal electricity technology. Sol Energy 74:19–31
73. Mir P (2012) Economía de la generación eléctrica solar. La regulación fotovoltaica y solar termoeléctrica en España. Civitas/Thomson Reuters, Madrid
74. Mir-Artigues P, del Río P (2016) The economics and policy of solar photovoltaics generation. Springer, Switzerland
75. Morales JI (2012) Las claves del éxito de la inversión en energías renovables. La transición de un modelo económico "energívoro" a un modelo económico sostenible. Instituto de Estudios Latinoamericanos/Universidad de Alcalá/Marcial Pons, Madrid
76. Naik H et al (2017) Medium temperature application of concentrated solar thermal technology: Indian perspective. Renew Sustain Energy Rev 76:369–378
77. Negro SO et al (2012) Why does renewable energy diffuse so slowly? A review of innovation system problems. Renew Sustain Energy Rev 16:3836–3846
78. NREL (2017) Concentrating solar power projects. Available at https://www.nrel.gov/csp/solarpaces/index.cfm. Accessed Mar 2017
79. OECD/IEA (2010) Technology roadmap. Concentrating solar power. International Energy Agency, Paris. Available at http://www.iea.org. Accessed May 2012

80. OECD, IEA (2017) World energy outlook 2017. International Energy Agency, Paris
81. Palenzuela P et al (2015) Concentrating solar power and desalination plants. Engineering and economics of coupling multi-effect distillation and solar plants. Springer, Berlin
82. Pariente-David S et al (2009) L'initiative de la Banque mondial en faveur d'une montée en puissance des centrales solaires à concentration (CSP). Utilisation du Fons des Technologies Propres en faveur d'une approche coopérative pou faire face aux défis du Changement climatique et aux défis énergétiques des pays du Moyen-Orient et d'Afrique du Nord (MENA). Réalités Industrielles, Nov 2009, pp 42–49
83. Perales T (2011) Instalaciones termosolares. Tecnologías, Sistemas y Aplicaciones. Creaciones Copyright, Madrid
84. Perlin J (1999) From space to earth. The story of solar electricity. Aatec Publications, Ann Arbor
85. Perlin J (2009) History of solar energy. In: Cleveland CJ (ed) Concise enciclopedia of history of energy. Elsevier, Amsterdam, pp 265–279
86. Peters M et al (2011) Shedding light on solar technologies-A techno-economic assessment and its policy implications. Energy Policy 39:6422–6439
87. Pietzcker RC et al (2014) Using the sun to decarbonize the power sector: the economic potential of photovoltaics and concentrating solar power. Appl Energy 135:704–720
88. Pitz-Pall R et al (2005) European concentrated solar thermal road-mapping. WP 3 deliverance no 7. Ecostar Team-DLR. Available at http://www.solarpaces.org/library. Accessed Apr 2010
89. Price H, Kearney D (1999) Parabolic-trough technology roadmap. A pathway for sustained commercial development of parabolic-trough technology. Report Number NREL/TP-550-24748. Available at http://library.umac.mo/ebooks/b12549289.pdf. Accessed Apr 2016
90. Pujol R (2009) Le retour d'un acteur français dans le secteur de la technologie du solaire à concentration. Réalités Industrielles, Nov 2009, pp 89–95
91. Purohit I, Purohit P (2017) Technical and economic potential of concentrating solar thermal power generation in India. Renew Sustain Energy Rev 78:648–667
92. Ragheb M (2014) Solar thermal power and energy storage historical perspective. Available at www.solarthermalworld.org. Accessed Jan 2017
93. REN21 (2011) Renewables 2011 global status report. REN21 Secretariat, Paris. Available at http://www.ren21.net. Accessed June 2012
94. Rivoire B (2005) Le solaire thermodynamique. In: Bobin JL, Huffer E, Nifenecker H (eds) L'énergie de demain. Techniques, environnement, économie. EDP Sciences-Grenoble Sciences Université Joseph Fourier, Grenoble, pp 371–400
95. Rosenberg N (1976) On technological expectations. Econom J 86:523–535
96. San Miguel G, Corona B (2018) Economic viability of concentrated solar power under different regulatory frameworks in Spain. Renew Sustain Energy Rev 91:205–218
97. Schinko T, Komendantova N (2016) De-risking investment into concentrated solar power in North Africa: impacts on the costs of electricity generation. Renew Energy 92:262–272
98. SAM (2018) System advisor model 2018 (version 2018/11/11). NREL, Golden, CO. Available at https://sam.nrel.gov/download. Accessed Dec 2018
99. Silvi C (2012) Italian contribution to CSP with flat or almost flat reflectors. Italian Group for the History of Solar Energy (GSES), Roma. Available at http://www.gses.it. Accessed May 2016
100. SL (Sargent and Lundy LLC Consulting Group) (2003) Assessment of parabolic trough and power tower solar technology cost and performance forecasts. NREL/SR-550-34440. National Renewable Energy Laboratory, Golden, CO. Available at www.nrel.gov. Accessed May 2009
101. Stadelmann M et al (2014) The role of public finance in CSP: Lessons learned. San Giorgio Group Policy Brief/Climate Policy Initiative. Available at http://www.climatepolicyinitiative.org. Accessed Dec 2016
102. Starnazzi C (2005) Leonardo. Codici & macchine. Florencia, Cartei & Bianchi editori
103. Teske S et al (2016) Solar thermal electricity global outlook 2016. Amsterdam. Available at https://doi.org/10.4321/S2254-28842015000200009. Accessed Nov 2018

104. Trieb F et al (2005) Concentrating solar power for the mediterranean region. Final report, German Aerospace Center, Institute of Technical Thermodynamics, Federal Ministry for the Environment, Nature Conservation and Nuclear Safety. Available at www.dlr.de. Accessed Apr 2010
105. Trieb F et al (2006) Trans-mediterranean interconnection for concentrating solar power. Final report. German Aerospace Center, Institute of Technical Thermodynamics, Federal Ministry for the Environment, Nature Conservation and Nuclear Safety. Available at www.dlr.de/tt/trans-csp. Accessed Apr 2010
106. Trieb F et al (2009) Global potential of concentrating solar power. In: SolarPaces conference, Berlin. Available at http://www.desertec.org. Accessed Apr 2010
107. Trieb F et al (2011) Financing concentrating solar power in the Middle East and North Africa—subsidy or investment? Energy Policy 39:307–317
108. Trieb et al (2012) Solar electricity imports from the Middle East and North Africa to Europe. Energy Policy 42:341–353
109. Trieb et al (2016) Rescuing the concept of solar electricity transfer from North Africa to Europe. Int J Energy Sect Manage 10(3):448–473
110. Trieb F, Müller-Steinhagen H (2009) The Desertec concept. Sustainable electricity and water for Europe, Middle East and North Africa. In: Desertec foundation, clean power from deserts. The desertec concept for energy, Water and climate security, 4th edn. White Book, Bonn. DLR/TREC/NERC, pp 25–46. Available at http://www.desertec.org. Accessed Apr 2010
111. Ummel K, Wheeler D (2008) Desert power: the economics of solar thermal electricity for Europe, North Africa, and the Middle East. Working paper, 156. Center for Global Development, Washington. Available at http://www.cgdev.org. Accessed Dec 2010
112. von Hirschhausen Ch (2010) Desarrollo de una súper red: conceptos y casos de estudio. In: Moselle B, Padilla J, Schmalensee R (eds) Electricidad verde. Energías renovables y sistema eléctrico. Marcial Pons, Madrid, pp 309–350
113. Viebahn P et al (2008) Final report on technical data, costs, and life cycle inventories of solar thermal power plants. Deliverable no 12.2-RS Ia, project no 502687 NEEDS/Sixth FP. DLR, CIEMAT. Available at https://www.solarthermalworld.org. Accessed Nov 2012
114. Viebahn P et al (2011) The potential role of concentrated solar power (CSP) in Africa and Europe—a dynamic assessment of technology development, cost development and life cycle inventories until 2050. Energy Policy 39:4420–4430
115. Vieira de Souza LE, Gilmanova Cavalcante AM (2017) Concentrated solar power deployment in emerging economies: the cases of China and Brazil. Renew Sustain Energy Rev 72:1094–1103
116. Williams TA et al (1997) A long-term strategic plan for development of solar thermal electric technology. SAND-97-1240C. Albuquerque, Sandia Corporation. Available at www.sandia.gov. Accessed Oct 2010
117. World Bank (2011) Middle East and North Africa region assessment of the local manufacturing potential for concentrated solar power (CSP) projects. Document available at http://arabworld.worldbank.org/content/dam/awi/pdf. Accessed Jan 2018
118. World Energy Council (2016) World Energy Resources. Available at http://www.worldenergy.org/wpcontent/uploads/2013/09/Complete_WER_2013_Survey.pdf. Accessed May 2018
119. Wyck JV, Lal AK (2010) FDI location drivers and risks in MENA. J Int Bus 9(2):99
120. Xu X et al (2016) Prospects and problems of concentrating solar power technologies for power generation in the desert regions. Renew Sustain Energy Rev 53:1106–1131

Chapter 4
Economics of Concentrating Solar Power Generation

4.1 The Value Chain of the Sector

The CSP value chain comprises many activities ranging from the development, civil works, solar field, tower, receiver, control, piping/valves, steam generation, turbine, cooling system, electrical system, auxiliary system, assembling, and research [15]. As of today, Europe is still the technological leader in the CSP sector and, given that one of the priorities of the Energy Union is to "become world leader in renewables", Europe is making efforts to preserve this status.

As demonstrated by various studies [6, 12, 19, 35, 67, 77], having industrial leadership brings multiple socioeconomic benefits in the form of employment and economic stimulation across many sectors. Besides the reduction of environmental externalities, the socioeconomic benefits of CSP deployment are important reasons that justify CSP support in many sunny belt countries.

As shown in Table 4.1, technology manufacturers along the CSP value chain are found in more than ten countries in Europe [15] and, out of the fourteen activities that comprise the CSP value chain, Spain ranks first, with a participation in thirteen of those.

However, and in line with what has happened in the wind power and PV sectors, the European leadership may quickly vanish due to the ambitious initiatives recently launched in other world regions and, in particular, China [47: 76–83, 81]. According to consulted experts, the growing threat on EU technology leadership comes from non-EU companies which have bought the industry's know-how holders and RD&D infrastructure at low cost.

Lilliestam [39] argues that the most critical aspects of the construction of a CSP project can be grouped into four main categories of the value chain: (i) the engineering, procurement and construction (EPC), (ii) the development of a project, including design and planning and also the plant components, (iii) the solar collector assemblies (the mirrors), and (iv) the receivers (heat collector elements, HCEs). Figure 4.1

P. Mir-Artigues et al., *The Economics and Policy of Concentrating Solar Power Generation*, Green Energy and Technology,
https://doi.org/10.1007/978-3-030-11938-6_4

Table 4.1 European participation in various activities of the CSP value chain

	Member state	Germany	Denmark	Czech	Spain	France	The Netherlands	Italy	Portugal	Belgium
Activity	Development	v			v	v		v		
	Civil works				v			v	v	
	Solar field	v			v	v				v
	Tower				v			v	v	
	Receiver	v	v		v		v			
	Storage	v			v					v
	Control	v	v		v	v	v	v	v	
	Piping/valves	v			v	v		v	v	v
	Steam generation	v	v		v		v	v		
	Turbine	v		v				v		
	Cooling system				v			v	v	
	Electrical system	v	v	v	v	v	v	v	v	
	Auxiliary system				v			v	v	
	Assembling				v			v	v	
	Research	v			v	v		v	v	

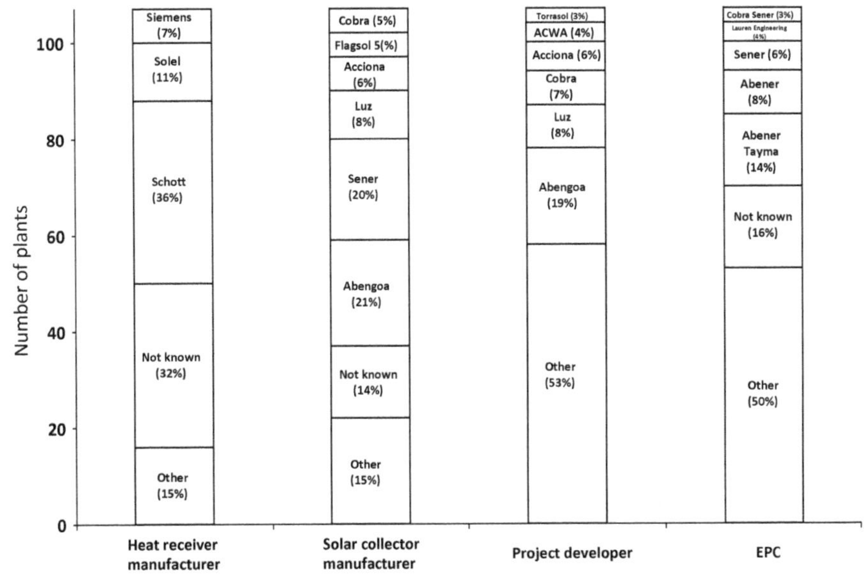

Fig. 4.1 Active companies in the EPC, developer, heat receiver, and solar collector manufacturer phases. *Source* Lilliestam [39]

shows the most important participating companies (for 107 CSP plants) for each of those categories.

To have a better understanding of the past, current situation, and future trends, a detailed breakdown of the country origin for the different supply chain phases is provided below[1]:

- As shown in Fig. 4.2, the EPC market is highly concentrated and dominated by Spanish companies. The market share of companies like Abener/Abengoa, SENER, Acciona, and Cobra (with 16, 14, 9, and 9 projects, respectively) is remarkable. However, since 2012, most Spanish EPCs have reduced their activity. In this regard, all but three projects under construction are built by EPCs without experience from previous projects. As highlighted by Lilliestam [39], whereas this is beneficial for new actors entering the relatively undiversified EPC market, there is also a considerable risk that the know-how acquired in the last decade is lost if the previously dominant companies exit the market [41].

[1]The data and figures presented here have been abstracted from [39], who conducted a very comprehensive analysis of a dataset of all CSP stations of 10 MW or larger in operation and under construction during the period 1984–2020 (csp.guru 2018).

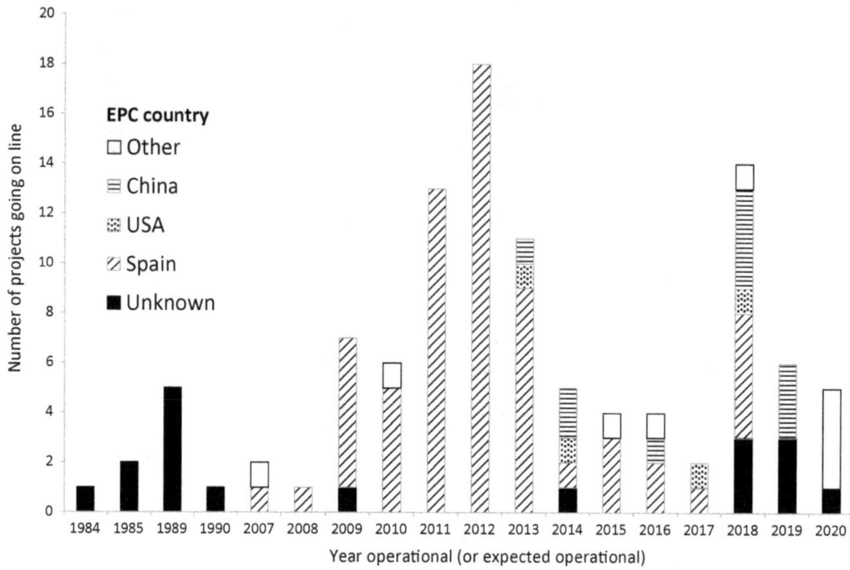

Fig. 4.2 Country of origin of the EPCs of CSP projects (1984–2020). *Source* Lilliestam [39]

- The market of developers is dominated by almost the same companies as in the EPC market, and as it can be observed in Fig. 4.3, After an initial leadership of American first and Spanish developers next, the market is currently experiencing an increase of Chinese and other Asian developers. As documented by Lilliestam [39], the Saudi developer ACWA has won bids for all three Noor stations in Morocco and for Bokpoort in South Africa and also won the bid for the 700 MW DEWA IV station in Dubai.
- In the market of heat receivers (HCE), Schott (bought by Rioglass in 2015) has dominated the market. Schott/Rioglass is active in at least two projects under construction, whereas all other known HCE suppliers are new (including, again, new Chinese market entrants) (Fig. 4.4).
- As in the EPC market, the production of solar collectors (SCAs) is dominated by the same companies which supply components to 60% of all projects. However, for the new projects, most SCA manufacturers are new. Contrary to the previous phases, the Chinese companies have not yet taken over the market [39] (Fig. 4.5).

The results from the analysis by [39] presented above seem to indicate that, after an initial industrial leadership of North America and Spain, new actors are entering the various phases of the CSP value chain. Among the newcomers, the emergence of Chinese companies is remarkable. According to [81], China's industry faces both great opportunities and challenges. The same authors argue that China has several positive features which could lead this country to achieve an industrial global leadership in CSP: It has large areas with excellent solar conditions for CSP, strong basic capabilities in traditional manufacturing activities which are important to CSP, and,

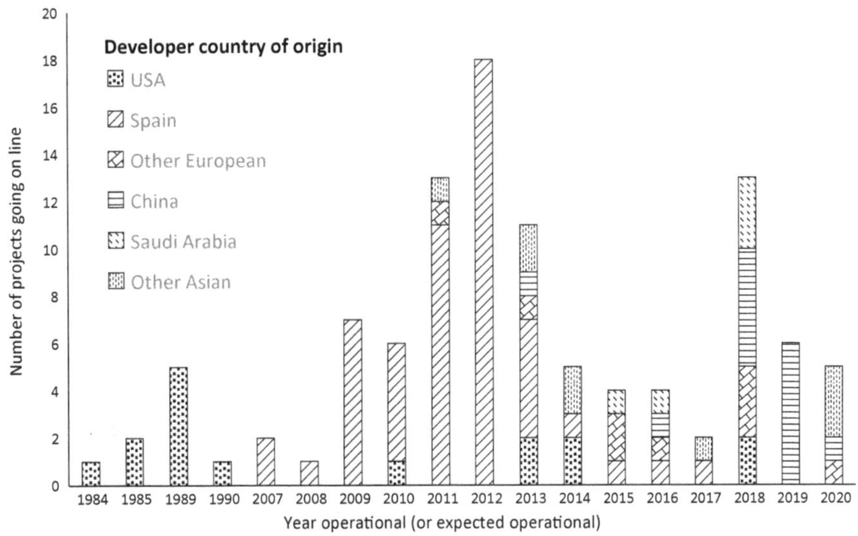

Fig. 4.3 Country of origin of the developer of CSP projects (1984–2020). *Source* Lilliestam [39]

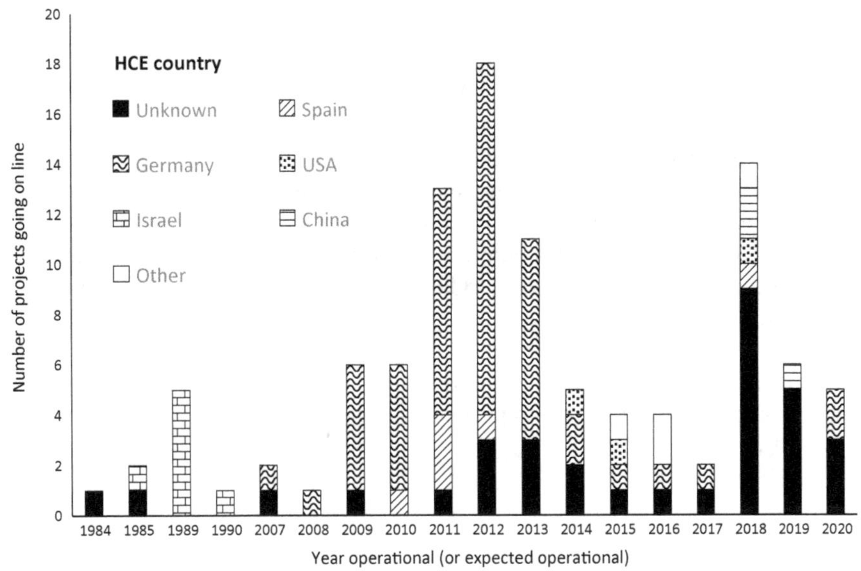

Fig. 4.4 Country of origin of the HCE manufacturer (1984–2020). *Source* Lilliestam [39]

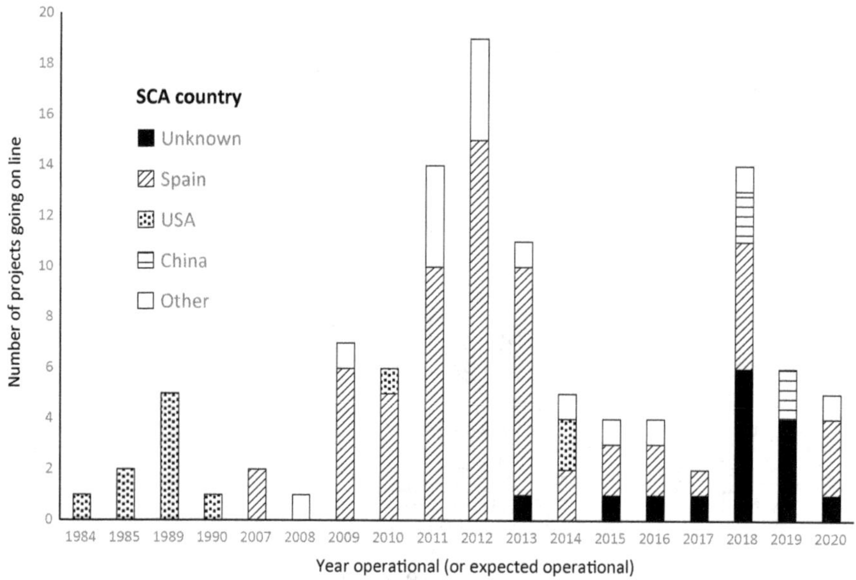

Fig. 4.5 Country of origin of the solar collector assembly manufacturers (1984–2020). *Source* Lilliestam [39]

also to some extent, know-how in CSP technologies. China would also profit from stronger international collaboration in the field, standardization, and international property rights legislation and management.

Regardless of the country of origin, the entry of new industrial players may have positive effects in terms of innovation as well as less dependence and vulnerability on the support scheme of a few countries. Furthermore, this will, in turn, contribute to ensure the continuity of the CSP industry in case the dominant firms leave the market [40]. According to the same authors, continuity in the industry is essential and support policies must be designed in order to address two risks: (i) Larger firms leave the market, and (ii) project developers and operators fail to take advantage of innovations and, as a result, fail to push costs down.

4.2 Design of Plants and Economic Analysis

The development of economic models for solar thermal generation, supported in their corresponding technical base, is an issue of some complexity for the following two reasons:

- Beyond the four main types of solar thermoelectric plants (i.e., parabolic trough, solar tower, linear Fresnel, and dish/stirling), there is a large number of particular configurations, each with their own technical and economic specificities. Some particularities are related to the resources of the place where they are located (DNI, water for cooling, etc.), to the goals of the public authorities in energy matters and to the willingness of the developers to improve previous designs (secondary innovations).
- The different energy transformations which were carried out in CSP plants are subject to several technical requirements which lead to numerous trade-offs, with the levels of energy efficiency being the key indicator of the quality of the process. The goal to reduce the cost of generation requires, then, carrying out very careful calculations. The issue is often to estimate to what extent the savings in one element (with respect to a reference point, probably a previous project) is not offset by the increase in the expenditures in another. However, such an increase may be a requirement in order to achieve the final reduction in the costs. It is not surprising that the process of economic optimization of CSP plants is arduous, given the dense network of technical and economic variables involved, as well as the uncertainty on the evolution of the later, which has kept generations of technicians and engineers busy since the middle of the nineteenth century.

The economic models proposed in this section are merely conceptual. Their aim is to highlight the basic technical and economic interrelationships which are present in the design process of thermo-solar plants, taking parabolic trough plants as the reference, since they are the most common design. There is no doubt that the identification of the technical and economic details of a specific plant is an issue which entails considering and fitting many relationships and optimizing a large amount of variables. Therefore, it is a process with many feedbacks.[2] Furthermore, it is likely that the definition of the project has to include the requirements suggested by planning and simulation models used by the system operators (SOs) in order to determine the least cost electricity dispatch generation mix (subject to given transmission grid constraints). Addressing such complexity goes well beyond the objective of these pages, although it inspires the models proposed. These models are built based on the following assumptions:

- The plant operates under a normal functioning regime (or steady-state conditions).
- The plant is only dedicated to the production of electricity. Complementary activities such as industrial steam production or water desalination are excluded.

[2]However, as it is obvious, the greater the experience of engineering firms, the greater the diligence in carrying out this activity.

- The plant has heat storage tanks and can operate under a hybridization regime.

In this context, two main issues will be highlighted: the discussion on the solar multiple (SM) concept and an approximation to the cost of generation per MWh.

Before addressing these issues, however, the economic process of a solar thermoelectric project should be briefly described. Although it is obvious that there might be many legal and financial variants, its main aspects are worth describing. To start with, given the large investment that a CSP plant represents (tens or hundreds of millions of € or $), a specific firm is created. There might be institutions which provide financial resources (such as economic and technology development agencies) to this firm. If it is a demonstration plant, public support is usually massive.

The next step is to contract an engineering firm which elaborates the first technical project and its business plan. These documents are then sent to financial institutions, which analyze the financial needs of the initiative and their risks and communicate their financial proposals to the firm, which will have to assess them. The accepted proposal is developed until there is a complete financial plan, which requires the approval of the financial institution, once third parties have revised the technical project and their regulatory and legal requirements.

If they are not shareholders, the financial institutions usually provide loans up to 70–80% of the funds needed, according to corporate finance or project finance schemes (the project itself is the collateral which secures the debt). In this last case, the cash flow coverage ratio is between 1.3 and 1.45 times the debt service. The amortization period of the loan is usually between 18 and 20 years. Sometimes, it is possible to cancel it after 7 or 8 years, with the shareholders assuming such liquidation, or the possibility to renegotiate the debt and its guarantees. Insurance companies cover unexpected events (delays in the execution, coverage of the loss of profit, and civil liability).

Next, the design of the definitive project, the setting up of the schedule for the completion of the project, and the subcontracting of the construction and purchase of the needed components and systems are awarded to a firm (probably through an invitation to a tender). They are engineering, procurement and construction (EPC) contracts, in which the main contracting party has to assume upward deviations, whereas it benefits from savings with respect to the awarded budget. Contractual formulas regarding the management of the plant by third parties during a trial period are also common. If the management firm anticipated part of the investment, this period may entail several years of operation until this investment is recovered. After such period, the management of the plant is transferred to its owners. O&M operations are usually subcontracted to a specialized firm, according to a fixed price which is periodically updated.

This institutional and financial scheme may be complicated with several issues: payment for the hiring of the land to the landowners where the plant is placed, bond issuance by the firm which owns the firm, etc.

4.2.1 Design Point and Solar Multiple

It should be mentioned first that solar thermal plants, as it is the case of other electricity generation plants, have to operate a maximum number of hours per year. Although there might be breakdowns and maintenance activities which lead to interruptions in their operation, CSP plants should operate during the night hours and when the impact of atmospheric circumstances reduces the DNI. Again, this leads to the storage of heat in order to operate at nighttime or to opt for hybridization. On the other hand, it should be mentioned that the capacity of thermoelectric generation grows with the intensity and persistence of direct solar irradiation (which represents 80–90% of the solar energy which impacts the earth crust). The minimum intensity values are between 1900 and 2100 kWh/m^2/year, whereas the persistence requires avoiding locations which are cloudy or have frequent mists, since those drastically reduce the DNI.[3] Although steam, aerosols, and ozone which are present in the atmosphere have a very small impact on the reflecting surfaces of the solar field, the dust carried away by the air requires their periodic cleaning [75: 58].

It should also be mentioned that the power block and the HTF/steam exchanger may operate in a wide range of partial load. This is an unthinkable attribute if the plant operates only under solar mode. However, for economic reasons, it should operate at its nominal (or full, or rated) power. This is why it may need the temporary recourse to the stored heat and/or hybridization. Furthermore, the number of daily hours of activity of the turbine changes depending on the season of the year. The more differentiated are the seasons, the more differentiated will be those activity intervals (assuming an only-solar without TES operation mode) (see Fig. 4.6).

With clearly differentiated seasons, the daily operation interval of the plant is greater in spring and summer, given the higher number of sunny hours. The opposite occurs during the winter and autumn. However, during the daytime hours, the operation of the power block is very stable and the closest possible to its nominal power.

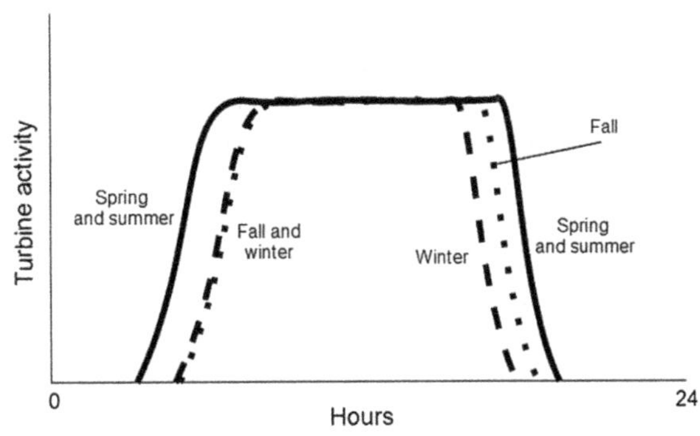

Fig. 4.6 Seasonal turbine activity. *Source* Own elaboration

[3]Once the clouds have gone by, it will take some time for the plants to recover their full level of activity.

As it is well known, the capacity factor (L) is a relevant indicator of the performance of an electricity generation process. Since generation can undergo interruptions (breakdowns, night hours in the case of solar plants) and oscillations (insufficient wind, cloud passing, etc.), the capacity factor indicates how many hours, taking the natural year as a reference, would have been needed if the plant had operated at full capacity, in order to generate the electricity that it really has generated. Or, in other words, in the case of CSP it indicates the equivalent amount of hours that the power block has been operating at full power in a year. The capacity factor is a technical indicator, although it has profound economic implications: The higher its value, the better the installed capacity will be used and the faster the investment will be recovered. The capacity factor of a solar thermal plant (defined without hybridization, i.e., only the solar generation) can be expressed as follows (adapted from Izquierdo et al. [29: 6216–6217]):

$$L = \frac{q}{\Lambda \cdot H}$$

where q is the electricity generated in a year (MWh), Λ is the nominal power of the turbine (MW) and H is the number of hours in a year (8760 h). As it was indicated, L is the relationship between the effective and the maximum generations.

In order to illustrate the underlying factors, this expression can be rewritten as follows:

$$L = \frac{\varepsilon \cdot \psi \cdot S}{\omega \cdot \Lambda_{\text{th}} \cdot H} \tag{4.1}$$

where ε ($0 < \varepsilon < 1$) denotes the collector system performance, Ψ refers to the solar direct irradiation captured by the solar field, S is the solar field collector surface,[4] ω represents the conversion factor between thermal energy and electricity,[5] and Λ_{th} is the capacity of the power block in thermal units. The numerator, thus, shows the quantity of energy delivered by the solar field (MWh), which is below the incident energy due to losses, whereas the denominator is the rated power cycle.

Leaving aside the technical details, the solar energy available is defined by the multiplication of the DNI and the collector surface, including losses. The latter depends on the type of plant. For example, in the case of a trough collector it is the aperture area of the parabolic reflectors, whereas, in a Fresnel plant, it is the overall surface of the flat mirrors. Inevitably, there are optical losses of the solar concentrator and receiver devices, as well as thermal losses of the HTF. The total addition of the losses of the solar field due to optical and geometrical reasons can represent more than 60% of the incident solar energy.

---

[4]The capturing surface of the collectors (S) is only a fraction of the total land area occupied by the plant. Izquierdo et al. [29] assumed that this is 27.5% for parabolic troughs and 12% for solar tower heliostats.

[5]This conversion factor is 1 kWh = 3.6×10^6 Joules (J), since 1 J = 1 W s.

A brief discussion on the relation between the thermal units (which are used to measure the capacity of the solar field, as well as the capacity of the power block) and the electrical units (which are much more common to measure the capacity of the power block) follows. The use of thermal units does not entail major challenges, although the different sources of heat which feed the power block, as well as unavoidable energy losses, have to be taken into account. Thus, given a sufficiently long time period (a year, for example), the thermal energy required in the process (E_{th}) is the addition of the energy provided by the solar field and used immediately $\left(\alpha_1 E_{\mathrm{th}}^F\right)$, or the energy charged/discharged by/from the storage system $\left(\alpha_2 E_{\mathrm{th}}^F\right)$, $\alpha_1 + \alpha_2 = 1$, and the one corresponding to hybridization $\left(E_{\mathrm{th}}^Y\right)$. Therefore, we can write:

$$q = \omega\left[\left(\alpha_1 \varepsilon \Lambda_{\mathrm{th}}^F + \alpha_2 \varphi^+ \varphi^- \Lambda_{\mathrm{th}}^F\right) H^F + \chi \Lambda_{\mathrm{th}}^Y H^Y\right]$$

where the electricity generated (q), given a conversion factor ω between thermal energy and electrical energy, is associated with the aforementioned thermal contributions. These result from the sum of the nominal thermal power of the solar field (Λ_{th}^Y) multiplied by its hours of activity (H^F), which is divided by a fraction α_1 immediately used and a part α_2 stored for a later use, plus the thermal rated power of the gas turbine (Λ_{th}^Y) which is used in a hybridization regime[6] multiplied by its hours of operation (H^Y). The right-hand side of the expression contains two additional parameters whose meaning is as follows:

- χ ($0 < \chi < 1$) represents the performance of the hybridization generation process.
- φ indicates the performance of the process of heat transfer from the working fluid to molten salts (φ^+) which is in the storage tanks, or the recovery from these tanks (φ^-). In both cases, yearly average values and $0 < \varphi < 1$ are considered.

The electricity self-consumption of the plant should also be taken into account. The annual electricity production, q, is a gross amount (MWh), since the plant consumes part of its own electricity given the needs of the pumps which operate in the plant, in order to feed the solar tracking devices and to maintain the cooling equipment active. Thus, in a parabolic trough plant, the HTF should be boosted within the solar field and the collectors should be moved to left/right. In the case of a solar tower, the heliostats have to strictly follow the transit of the sun. This is also the case with the dish/stirling collectors. As a result, the plant usually consumes between 5 and 10% of the electricity that it generates. Thus, differently from other renewable technologies, CSP plants consume a non-negligible amount of electricity when they are in operation. Thus, the annual quantity of electricity fed into the grid (MWh), which is remunerated at a given price p, is $q^* < q$.

Assuming the existence of an energy policy which promotes thermo-solar generation, as well as a sufficiently detailed plan for the deployment of new-generation

[6]Assuming that the plant is hybridized with natural gas simplifies the expression. Indeed, it is directly burned in the corresponding turbine whereas, if the aim is to use the gases from coal or biomass combustion, they should be channeled to a heat exchanger in order to obtain superheated steam.

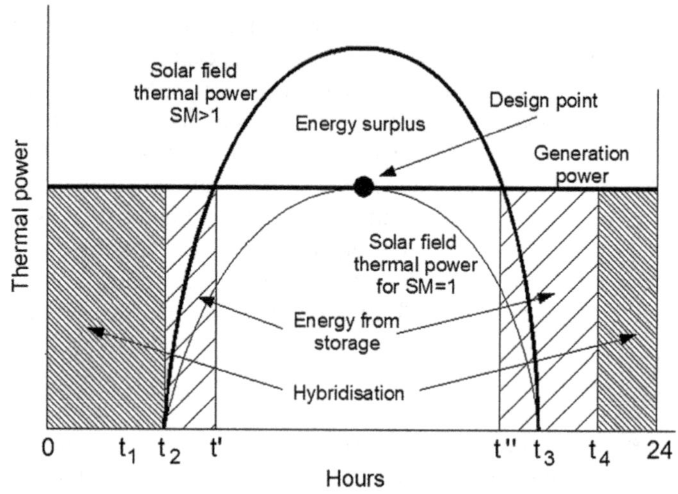

Fig. 4.7 Design point, solar multiple, storage, and hybridization. *Source* Own elaboration

capacity and grids, the first step of project developers of a new CSP plant, within the regulatory framework, is to set the electricity capacity, that is, the nominal capacity of the power block. These data are essential for the later design of the solar field and the rest of auxiliary systems of the plant.

The goal is to adjust the thermal capacity of the solar field to the thermal needs of the power block, with the aim to sustain its electricity generation capacity. However, the thermal energy delivered by the solar field is not constant, and there are hours (and even days, depending on the latitude), in which this can be lower than the required thermal power of the turbine. In other moments, the volume of thermal energy exceeds the needs of the power block. In order to solve this mismatch, a first step is to calculate the so-called design point or operating point under steady-state conditions, that is, the size of the solar field which delivers a sufficient amount of thermal energy to run the power block. Figure 4.7 (based on Palenzuela et al. [56: 106]) illustrates this discussion.[7]

The design point is obtained by considering a wide array of factors, including the following for a parabolic trough: the orientation of the axis of the collectors, the geographical location of the plant, with a special attention on the irradiation and climate of the place, the incidence angle of the direct solar irradiation on the collectors, the difference of the temperature of the HTF when it enters and leaves the solar field, the type of collector, and the type of working fluid and its optical and thermal losses, respectively [56: 92–106]. All this has to be adjusted to the incident irradiation in a given moment of the year, for example, the one corresponding to the summer solstice at noon. With all this, the size and technical features of the solar

[7]The asymmetry of the figure is worth noting: The energy stored is higher in the afternoon than in the morning.

field which allow meeting the thermal power required by the generation system are obtained. This is the so-called solar multiple (SM) which takes the value of 1 (in Fig. 4.7 SM = 1 curve).

Generically, the SM is the existing relationship between the nominal thermal power collected by the solar field and the surface necessary for the turbine to work at its rated power or, also, its needs of thermal input [7: 682, 25: 14, 29: 6215]. Therefore, according to expression (4.1),

$$SM = \frac{\psi \cdot S}{\Lambda_{\text{th}} \cdot H} = \frac{\omega}{\varepsilon} \cdot L$$

The SM is closely related to the capacity factor. If SM increases, the capacity factor will also increase, taking into account that the level of energy losses of the plant and the thermal energy/electricity conversion factor do not change.

In Fig. 4.2, the thermal power corresponding to SM = 1 is only the starting point in order to expand the size of the solar field. SM = 1 corresponds to the sizing of the solar field so that, then, it is enough to replicate the technical unit, thus obtained in order to meet the thermal requirements of the power block only by using the energy from the solar field, during the annual number of hours which the managers of the plant deem appropriate (see the SM > 1 curve of Fig. 4.7). This resizing of the solar field allows meeting the thermal requirements of the power block during the daytime hours. This barely changes if the plant is in the appropriate latitudes (see Chap. 1). Otherwise, the seasonal variations in the solar irradiation need to be carefully considered and offset through other ways to generate steam. Setting a SM > 1, however, leads to an excess of thermal energy in the middle hours of the day, as shown in Fig. 4.7. Since the alternative to change the focus of some collectors does not have economic sense, another solution will have to be looked for.

The value of SM is between 1.1 and 1.5 in plants without heat storage or hybridization. In this case, all the energy captured by the solar field is transformed to electricity although, as it is obvious, this may lead to periods in which the turbine/generator operates below its nominal capacity. The activity of the plant will be null at night and cloudy days. The economic efficiency of the generation process is, thus, seriously jeopardized. In order to improve the capacity factor, the excess heat provided by the solar field, that is, the volume of thermal energy which is above the needs of the power block, can be stored and converted into steam at night hours. With an SM which has values in the range between 2 and 4, the plant is close to its objective to operate at its rated power the maximum number of hours in a day and the maximum number of days in a year. The surplus of thermal energy provided by the solar field is not a problem, but rather the opposite is true: It allows increasing the capacity factor and, then, the economic efficiency of the plant. The fact that storing heat is relatively cheap (compared to storing the electricity directly) allows spreading out the size of the solar field, despite the additional investment that it requires,[8] since it

[8]As it is obvious, increasing the SM value also increases the land area required by the plant. If a plant is located in an arid zone, then this aspect is irrelevant.

is offset by the greater capacity factor [72: 8–11]. The storage achieved by adding solar fields with a size of SM = 1 avoids having to change the focus of the collectors in case of surplus. It also allows extending the hours of electricity generation since there is sufficient thermal energy. If storing heat had been an expensive operation, CSP generation would never have taken off, as warned by some of its pioneers (see Sect. 3.1).[9]

In reality, the process of adjustment of the size of the solar field, the capacity of the power block, and the hours of storage are carried out through simulations whose output is the cost of the MWh generated by the plant. The accumulated experience suggests that, as a general rule, the optimal values for the SM are between 2.5 and 3, both for parabolic trough plants and for solar towers, whereas the hours of thermal storage are usually 4, 8, 12, or 16 h. The greater the number of hours of TES which are added to the indicated SM values, the lower is the LCOE, although the advantage from 8/10 h of storage is negligible [50, 51: 62]. On the other hand, the capacity factor increases with the value of SM and the hours of storage, although proportionally less with a higher number of hours of storage. This result clashes with the increase of the investment that the heat tanks involve. At the start of the current decade, the simulations carried out indicated that, in the case of parabolic trough plants, the lowest costs (i.e., €cts 16/kWh) were associated with a multiple of 2.5 and 8 h of storage. For solar towers, the SM has a secondary role, however. The lowest cost (€cts 10/kWh) corresponds to the combination of SM = 4 and 16 h of storage, although for SM = 2.5 and 8 h of storage, the cost of the MWh only increased by 1 €cent. It should be noted that, in principle, the solar towers had costs of the MWh which were comparatively lower than those of the parabolic trough plants, whatever the number of storage hours (see Izquierdo et al. [29: 6219–6221], Jorgenson et al. [31, 32], Mehos et al. [45]).

Figure 4.7 also shows that the storage capacity, measured as hours of operation of the power block at its nominal power, does not have to be used in a continuous manner. Thus, it has been assumed in the figure that, after sunset, the plant uses the heat accumulated during the day, a fact which allows it to meet the likely high electricity demand at sunsets. Notwithstanding, when the night advances, and assuming a limited storage capacity, the plant uses generation by hybridization (with its alternative being to stop its activity of the power block). At dawn, the stored heat is used again in order to guarantee a normal operation of the power block, while the DNI increases as the sun goes up in the sky. It should not be forgotten that the losses of the tanks are negligible and, thus, the suggested time distribution of use is perfectly possible. Finally, when the quantity of heat collected in the solar field exceeds the capacity of the thermodynamic cycle, part of the fluid is deviated to the TES, where it is stored until sunset. It has been assumed that after both lapses of activity from the storage (the one at sunset and the first daytime hours), its capacity has gone down

[9]The positive economic effect of storing heat was taken into account at the start of the twentieth century (see Chap. 3). It was clear at the end of the past decade that the increase in the investment per installed MW, related to the increase in the fraction of solar in generation, would lead to higher capacity factors and, then, to a gradual reduction in generation costs [76: 125].

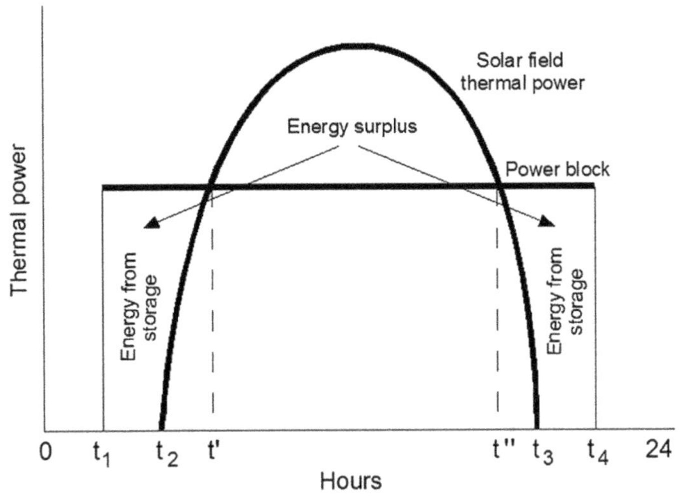

Fig. 4.8 Thermal energy surplus in daily central hours of the day. *Source* Own elaboration

to the minimum level, and thus the surplus which is accumulated will allow a new activity cycle.

It should be mentioned that the analysis carried out so far does not contain any assumption about the use of the plant. This use is associated with the desired time lapse of the activity. Let us start from Fig. 4.8, which represents an already known situation: There is an interval of daytime hours $[t', t'']$ in which the solar field provides more thermal energy than needed by the power block. This surplus is stored in order to be able to generate electricity when there is not any solar light, i.e., in the intervals $[t_3, t_4]$ and $[t_1, t_2]$ or, which is more common, adding to these intervals those in which the irradiation being captured is not enough anymore to feed the turbine, that is, $[t''$, $t_4]$ and $[t_1, t']$. This regime of operation does not rely on hybridization, as shown in the figure.

However, the case shown in Fig. 4.9, albeit opposite to the previous situation, could also occur since it makes economic sense. There could be a plant whose only objective is to cover the electricity demand in hours with high consumption, that is, at midday in hot days. Therefore, the heat produced is stored in the first and last hours of the day in order to reinforce electricity generation at times of peak demand. In this case, the thermal needs of the power block are above the maximum direct thermal contribution of the solar field. In order to avoid an oversized generator with respect to the solar field, the heat produced in off-peak hours is stored. Of course, a careful calculation has been required in order to adjust the surplus and deficit of thermal energy. As it is obvious, the profitability of this design assumes a high price of electricity in the midday hours (demand peak load) [25: 14–15]. It all seems to indicate, however, that this possibility has vanished due to the competition with photovoltaic generation.

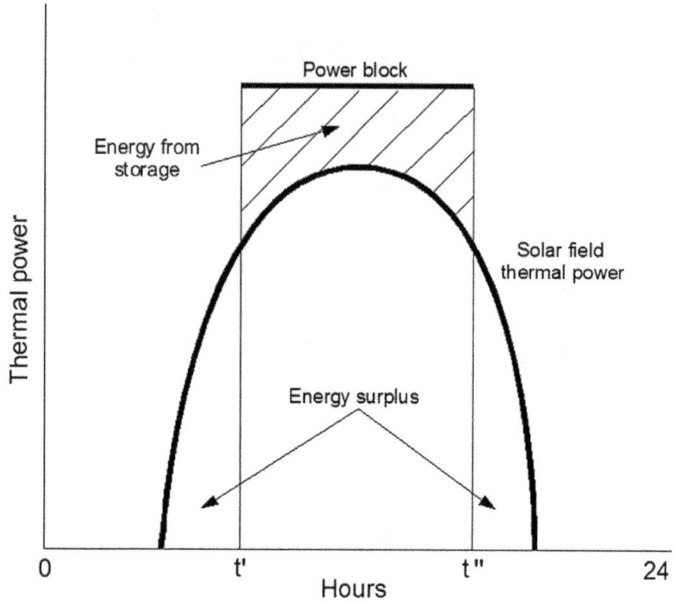

Fig. 4.9 Thermal energy deficit in the central hours of the day. *Source* Own elaboration

Therefore, the production of a CSP plant is governed by the SM. The higher the value of the solar multiple, the higher is the capacity factor. However, this implies a higher investment and availability of land (a possible limiting factor, which is much less relevant in arid zones with even land). Fortunately, the heat captured by the solar field which is above the needs of the power block can be stored at a reasonable investment cost and a low operation cost, with negligible losses. Therefore, the number of hours of operation of the power block can be extended, which distributes the weight of the investments among a much larger number of MWh. In reality, this offsets the problem of having installed a disproportionate solar field and the TES (which will feed the power block beyond the daytime hours). Thus, with 8 h of storage or more, the capacity factor may reach values above 60% [28: 84], which doubles the capacity factor without storage.[10] The result is a lower-generation cost per MWh. Some studies carried out in the past show that this cost progressively goes down with the increase in the SM and the storage hours until a minimum stretch is reached, which is very similar for slightly different plant configurations [29]. Obviously, technological changes as well as economic incentives can lower the generation cost even further. Indeed, the other variable which affects the economic performance of a plant is the advantage and requirement provided by regulation. Energy policy and its implementation can set up preferential tariffs, tenders, fiscal incentives and subsidies, limits to hybridization, etc (see Chap. 6).

[10]Hybridization is the alternative. However, it is subject to the evolution of the prices of the fuels being used.

4.2.2 Economic Analysis

Without losing sight of the considerations made in the previous section, the total annual cost (C_t) of electricity generation by a CSP plant in a given year can be expressed as follows

$$C_t = \left(I + I^F + I^S + I^Y\right)\frac{i(1+i)^T}{(1+i)^T - 1} + \overline{W} + \Lambda m_t + \langle \Lambda H f p\rangle^Y$$

where

I	refers to the investment for the purchase and installation of fixed fund elements of the process,[11] such as the power block, the tower (in case of a plant with a central receiver) and the buildings, and auxiliary equipments, expressed in monetary units (€, $, etc.).
I^S, I^F, and I^Y	represent the investment in the TES equipment, the solar field, and the hybridization system, respectively, expressed in monetary units.
$\overline{W}$	represents the annual payments for the services, or wages, of the human work fund involved in the control of the generation process.
m_t	is the annual O&M costs, expressed in monetary units per MW.
T	refers to the operational lifetime of the plant which, for the sake of simplicity, is assumed equal for its different equipments and systems.
I	is the interest rate applied in calculating the depreciation annuities.

As it can be observed, the investments in the solar field (trough rows, heliostats, or dish), in the TES, and in the hybridization systems (adding the equipment purchase and the setting up operations) are considered isolated from each other due to their particularities. The expenditures incurred in the elaboration and administrative processing of the project have also been added to the amount of investment. With respect to the annual wages, they probably are a comparatively small amount. Regarding O&M costs, it is assumed for simplicity reasons that they include the annual costs of the different flows needed for solar generation, such as the lubricants and spare parts. The term on the extreme right represents hybridization: the annual MWh which is generated by consuming natural gas (f^Y per MWh) at a price p^Y. Finally, the cost of purchase or hiring of the land has been ignored in the expression, and it also leaves aside the possible incentives in the form of subsidies, fiscal reductions, etc.

If C_t is divided by q, the cost per MWh_{AC} is obtained (€/MWh_{AC}). It should be indicated that part of q, or electricity generated in a year by the plant, is produced but not sold; i.e., it is self-consumed.

The different components of the equation of the cost have a very different weight. For illustrative purposes, if a 95% of efficiency in the thermal storage, an interest rate of 9%, and a useful lifetime of the plant of 25 years are assumed, the values of the

[11]For the definition on the fund and flow elements in a production process, see Mir-Artigues and González-Calvet [48].

Table 4.2 Economic magnitudes (2005[a])

Variable	Unity	Parabolic trough	Solar tower
I	€/W	1.37	2.05
I^S	€/kWh	90	40
I^F	€/m^2 collector	213	150
m_t	€/kWh/year	0.12	0.146

[a]Yearly average value 2005: €1 = $1.24

Table 4.3 Economic magnitudes (2015)

Variable	Unity	Parabolic trough	Solar tower
I	€/W	0.87	1.45
I^S	€/kWh	69	28
I^F	€/m^2 collector	200	144
m_t	€/kWh/year	0.02–0.03	0.03–0.04

Source Own elaboration from Mehos et al. [44: 31–32] and IRENA [28: 8], and experts' advice provided to the authors

main variables, for the year 2005, would be those indicated in Table 4.2 (as shown in Izquierdo et al. [29: 6217]).

On the other hand, Table 4.3 shows more recent data.

In addition to the generally observed reduction, the interpretation of the numbers in both tables should take into account that they do not correspond to a specific plant. They only represent indicative values.

If the data presented are extended, a possible detailed disaggregation of the investment in a parabolic trough plant (based on Stoddard et al. [10: 22, 74: 5–5]) can be as follows:

- Around 50% is accounted by the solar field, with at least half of this percentage corresponding to the mirror support structured and the mounting. The absorption tubes as well as the HTF storage tanks also stand out.
- The power block itself does not reach 10%, although if control and firefighting system installations are added, the percentage can increase to between 10 and 15%.
- The storage system (tanks and heat exchangers) represents about 20%.
- The electricity installations represent between 5 and 10%.
- The rest of the investment corresponds to civil works, engineering, and administrative processing.

In the case of a solar tower, the numbers are similar, with the logical exception of the heliostats, which represent more than 1/3 of the total investment, and the central receiver (more than 10% of the total). Therefore, for both technologies, the cost of the solar field, the storage systems, and the power block is above 4/5 of the total investment.

Regarding O&M costs, Stoddard et al. [74: 5–5] indicate that labor (32%), repairs, and spare parts of the solar field (28%) and the rest of systems of the plant (10%) stand out.

Table 4.4 Efficiency criteria for CSP plants (2010 and 2015)

Collector type and turbine	Sun concentration, and peak and annual solar efficiency	kWh/year per m^2 of occupied land	Thermal cycle efficiency (%)	Annual solar-to-electricity conversion (%)		Capacity factor (no TES)
				2010[a]	2015[a]	
Trough, steam turbine	80, 21%, 17–18%	45–55	30–40	11–16	15–16	~25%
Tower, steam turbine	300–1000, 35%, 25%	70–90	30–40	12–16	15–17	
Tower, combined cycle			45–55	20–25	–	
Fresnel, steam turbine	25–100, 20%, 12%	50–60	30–40	8–12	8–10	
Dish	1000–3000, 30%, 24%	80–100	30–40	15–25	–	

[a]Year in which data were published

Regarding the evolution of the amount of investment, a 100 MW plant in 2008 required about 4900 $/kW of investment [58: 44]. This number was expected to increase in the short term, both for parabolic trough and for solar towers, due to the addition of TES with gradually more storage hours. However, given a capacity of the TES of 14 h, some projections indicated a reduction in costs to a minimum of $3000/kW for the decade of 2030 [14: 10-25/10-28].[12]

As a complement to the tables above, Table 4.4 shows the values of the efficiency indicators which are most common for thermo-solar plants. Those indicators have been:

- Sun concentration and the peak and annual solar efficiency.
- The electricity generated by the surface occupied by the plant and the collectors.
- The relationship between the direct irradiation captured by the collectors and the electricity generated by the plant, or solar-to-electricity efficiency, per unit of time (a year, for example).

Given its capacity to concentrate the sun rays, the dish/stirling and the solar tower stand out, although the greatest land-use requirements correspond to the Fresnel technology [7]. A solar tower generates less electricity per unit of land due to the large quantity of land required by a field of heliostats. However, these plants have a more homogenous generation profile throughout the year, since the heliostats are

[12]In this section, data on the LCOE (MWh) are not included. See Chap. 3 and [28].

always perfectly oriented to the sun and their annual solar efficiency is the highest. Whereas the capacity factor is very similar for all the plants, the values of thermal efficiency as well as the solar-to-electricity conversion factor are quite different. In this last case, the table shows the values for the end of the past decade and the middle of the present decade [2: 1009, 1012 and 1017].

4.3 The Values of Concentrating Solar Power Electricity Generation in a Changing Electricity System

The main impacts of CSP electricity on the current social, economic, and energy context have to be systematized. Therefore, first, the stages of the transformation of the electricity sector due to the progressive penetration of renewable electricity sources are described in a stylized manner. Then, the role of CSP generation is placed in such a context. Whereas the first issue is analyzed for the first time, the second one follows and expands the systematization proposed by Mir-Artigues and del Río [47: 113–152].

The goal of this section is, thus, to identify the main role of CSP generation in the different stages of the structural change of the electricity sector. In other words, the aim is to determine the value of CSP generation, which is understood as its contribution to the success of the energy transition, that is, to the evolution toward an electricity system with a dominant role of renewable energy sources. This contribution results from the combination of technical, economic, and social features which are ideal to encourage such change, as well as features whose effects hinder it and which should be mitigated in one way or another. Given that, at least today, four renewable energy technologies (wind power, PV, CSP, and biomass plants) compete to be the main agent of this transition, the next pages, which discuss the pros and cons of one of them (CSP), provide only a generic diagnostic. Therefore, it is not directly applicable to specific national electricity systems given their different electricity generation mix and particular socioeconomic requirements.

4.3.1 A Stylized Model of Structural Change of the Electricity Sector

A key economic objective of renewable electricity support is to create the conditions to reduce the cost of generation (€/MWh) to levels comparable with conventional energy sources in a reasonable time span. This is an economic requirement in order to advance toward the greatest possible decarbonization of the electricity sector. Reaching a sustainable electricity sector is, however, a complex process [17: 1175–1201, 62, 73]. This entails a long transition in which the generation mix gradually changes, new roles for T&D networks appear, new markets emerge, etc. All these happens at the same time electricity demand and stability and reliability of the electricity system are guaranteed.

A stylized model related to the structural change process [59] which is being experienced by the electricity sector is provided. This process has been divided into four stages, as shown in Fig. 4.10. This is a representation of the behavior of the main economic variables involved in the transition from an electricity generation mix dominated by technologies which emit GHG (use of hydrocarbons and coal as fuels) and nuclear generation rejected in many places due to the risks that it entails, to a mix in which renewable energy sources and, among them CSP, dominate. Obviously, the figure represents only a hypothetical conceptual framework. Its objective is to highlight the alleged evolution of costs and prices over time. In this sense, it should be taken into account that

- The figure does not have a timescale. The lapse of time covered by each stage includes a very different number of years depending on the country or region, whereas the whole process can be extended for decades.
- The factors determining the relative positions of the economic variables which have been considered are more relevant than the trajectories of those variables.
- The transition process entails gradual changes in the generation mix and, thus, the concomitant adjustment of T&D grids. Each country would represent a particular case because of the historical path being followed, but also due to the presence of the cheap primary sources in such country. This is not considered in this model: The proposed framework only pays attention to the consequences of those changes on costs and prices.
- The economic values are expressed in real terms.
- The configuration of the later stages to the present moment, that is, the end of the second stage onward, is merely speculative. The figure does not aim to be a prediction: The lines corresponding to the last two stages of the structural change process reflect the predominant expectation, although not unanimous, among the experts. There is a great uncertainty in the analysis.

To start with, p denotes the preferential tariff (whether a feed-in tariff, premium, or green certificate), c is the cost of renewable energy generation (whose rate of reduction has been assumed gradual), e is the retail price of electricity (taking into account that, although not all consumers pay the same amount per kWh, the tariffs move in tandem, i.e., they share a trend), and w represents the wholesale market price of electricity. Secondly, there are two prominent positions in Fig. 4.10:

- The points called *rpg*, which indicates the retail grid parity, and *wsg* which indicates wholesale price parity. Although they have been represented as points, they are really regions, since parity depends on many factors, some of them being idiosyncratic [47: 109–113]. It should be taken into account that not all renewable technologies achieve those parities at the same time. Obviously, if any accumulates a severe delay, its diffusion possibilities are negatively affected. However, they are not removed since there are many more factors at play than costs. In principle, however, the technologies which are deployed earlier have a greater chance to

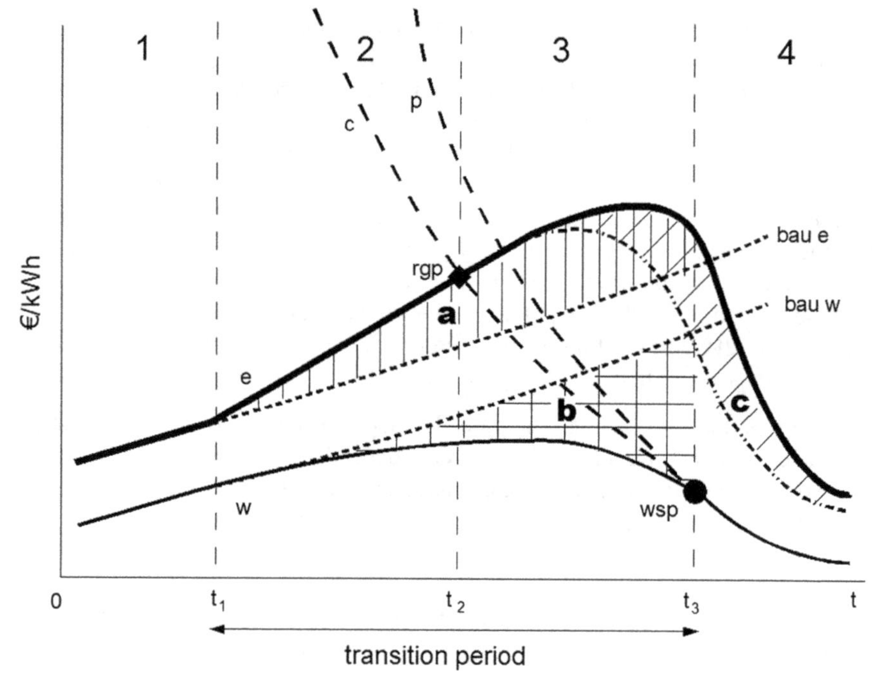

Fig. 4.10 Stylized stages of the energy transition process. *Source* Own elaboration

become dominant. It can be observed that line *p* is high above the cost line *c* up to the point *rgp*. Then, they tend to converge (which occurs in *wsp*). The reason for this behavior is that, with the gradual reduction in the costs of the renewable energy technologies, many regulations abandon the preferential prices (as well as other advantages). A premium is implemented which, when deemed necessary, is added to the wholesale electricity price. Indeed, the amount of the premium goes down (year after year, for example) given the reduction in the aforementioned costs, although there might be time spans in which the wholesale electricity prices are too low for renewable energy producers. At a certain moment, the premium will no longer be needed. Since then, the market prices already guarantee a profit.

- The shadowed zones *a*, *b*, and *c* represent, respectively, the additional increase of retail prices which is needed in order to finance the renewable promotion policy, the downward pressure of average wholesale prices due to the zero price at which renewables are offered, and the costs of providing backup to an electricity system with a strong presence of variable renewable energy sources (see below).

The upper solid line represents the hypothetical trend of the retail or final electricity price: The promotion of renewable energy sources leads to an increase above their historical trend in the case that the generation mix keeps on being only a conventional one. The circumstance that we would like to describe is that, even if a sustained increase in the price of uranium (the nuclear fuel) and hydrocarbons is assumed (including the impact of an eventual carbon tax on them), the financing needs of renewable sources, which initially have a very high generation cost, put further upward pressure on retail prices.

The consumers are the ones who bear the costs of the promotion scheme, although the regulation may distribute the costs in an unequal manner.[13] The magnitude of the increase experienced by electricity prices depends on the calculation of the remuneration and the trend in generation costs [11]. This is a problem that, with a high probability, will not affect those countries which promote renewables later on.

The solid but thinner line below indicates the particular trajectory of the wholesale electricity price with an increasing penetration of renewable electricity in the market. Since this electricity enters at a zero price in the market, the number of conventional plants which offer a higher price and, thus, are displaced increases. This impact on the merit order is greater than the amount of renewable energy that enters the market. Therefore, the trend of the wholesale electricity price gradually diverges from the trend which it would have followed without such penetration.

The displacement of conventional electricity generation plants accelerates if the costs of renewable energy technologies keep on going down. In the point noted as *wsp*, renewable energy sources with no premium start to be competitive in the wholesale electricity market.[14] At the end, the expectation is that wholesale electricity prices go down, which then ends up driving down the retail prices (FITs and FIPs fell behind), even taking into account the expenditures of backup generation (whether conventional or renewable[15]) and the financing needs of the grids.

The following subsection provides more details on the four stages of the structural change of the electricity sector shown in Fig. 4.10.

4.3.1.1 The Electricity Sector Is Still Conventional

In the initial stage, lapse $(0, t_1)$, there are only conventional electricity generation plants. These are huge hydro plants, as well as thermal plants which burn fossil fuels or nuclear plants which have large economies of scale. There is a complex T&D network which brings the electricity to the final consumers, whose role is merely passive. In many countries, after World War II, the electricity sector was dominated by a large vertically integrated public company although, in other countries, a reduced number of private companies kept operating. However, in the last quarter of the twentieth

[13] In case it is budget-financed, the analysis does not substantially change because the set of taxpayers and consumers coincide, given that electricity is a basic good in the sense of Sraffa [37].

[14] Furthermore, in addition to the displacement of the more expensive techniques, there might be a lower demand volume. This is due to two reasons: the modular character of renewable energy techniques, such as PV and mini-wind, which would allow the massive diffusion of prosumers (who may have storage systems [70], and efficiency and energy-saving policies, which would have an impact on such demand.

[15] Renewable energy dispatchable plants include biomass and solar thermal plants, although variable plants may participate in the intraday and balancing market, as well as in ancillary services. The electricity storage systems at scale and low cost will eventually reduce the required backup capacity.

century, liberalization processes were promoted, together with other privatization and concentration processes which led to generalized oligopolies, with different particularities depending on the country (see [18]). Simultaneously, agencies for the regulation of the electricity sector were created or encouraged.

In this stage, the discontinuous lines *bau e* and *bau w* start. They represent, respectively, the business-as-usual trends of final and wholesale electricity prices. It is assumed that both have an upward trend due to the impact of the increasing prices of hydrocarbons and the nuclear fuel, in a context in which neither renewable technologies nor the policies, which support them, are present. These trajectories reflect the gradual exhaustion of those primary energy sources,[16] although they are also related to the difficulty to exploit economies of scale in conventional electricity generation [42: 11–37]. Although some innovations, such as the ones which have improved crude oil extraction in bituminous sands or fracking, have allowed the exploitation of previously inaccessible wells, they have also delayed the concern about the scarcity of hydrocarbons (never an issue in the case of coal) but they have not stopped the concern about climate change [24]. The pressures from vested interests and/or the institutional inertia have not stopped the idea that it is necessary to advance toward a different energy model for environmental reasons.[17] The energy transition starts, aside from the prices of conventional fuels.

4.3.1.2 First Stage of the Transition Period

The change of the electricity sector starts in the period (t_1, t_2). In the beginning, the measures to promote renewable energy sources for electricity generation are shy and, thus, their presence of these sources in the system is rather residual, although they gradually increase. In historical terms, this stage started in the last third of the last century: first between the mid-1970s and the 1980s in USA as the main pioneer and, then, since the 1990s with Germany and Japan as leaders. Initially, the concern about the exhaustion of fossil fuels dominated. Then, climate change became a major concern.

Initially, some were reluctant on the need to support renewable energy technologies. In addition, the choice of the most suitable promotion scheme and its detailed design were issues solved through the essay and error method. A key decision had to be taken: whether to support investment paying preferential prices for the kWh (demand-pull option) or support the supply side by favoring the RD&D expenditures (or technology-push option). The dilemma was solved by simultaneously activating both options [54], although pioneer countries focused more on one or the other. In fact, the debate on the need to prioritize either of the two options and, also, about the specific type of demand-pull measure to implement was alive throughout the period.

[16] In some countries, this is aggravated by the dependency on third parties, since it may involve supply problems due to political reasons.

[17] For this reason, the argument does not change if the *bau e* and *bau w* curves are assumed to be horizontal or slightly declining.

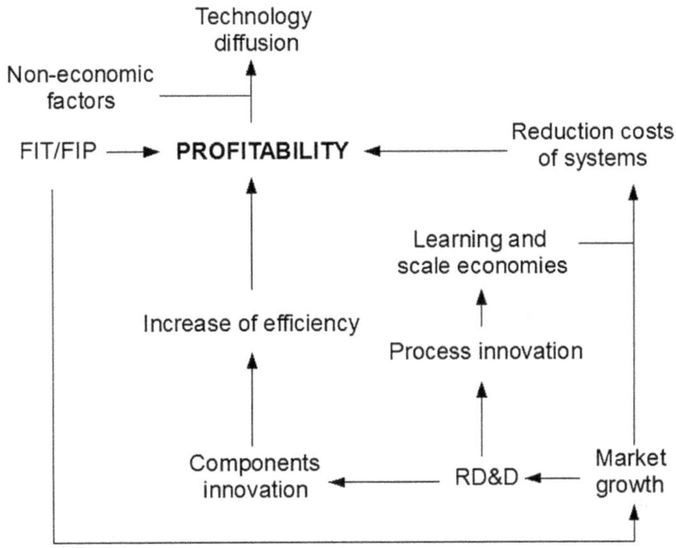

Fig. 4.11 Economic rationale for supporting renewables. *Source* Own elaboration

Both measures were arranged within a given economic rationale, whose representation is illustrated in Fig. 4.11.

The proposed scheme shows the connection between the variables which, presumably, would guarantee the growing diffusion of the renewable energy technologies by lowering the generation costs. The interpretation of the figure may start from the technical and scientific knowledge accumulated after World War II and, specially, between the 1970s and the first half of the 1980s, although its roots go back to the nineteenth century, as it is the case with wind, PV, and CSP ([47, 52]: Chap. 3 and Sect. 3.1 of this book). The knowledge accumulated in RD&D activities encourages improvements in the efficiency of components (solar collectors, cells and modules, rotors, power blocks, and so on)[18] and process (directed to reduce the manufacturing costs of the different components). Furthermore, they encourage the adaptation of innovations from other fields (or spillovers). However, renewable energy technologies are initially very far from the competitiveness frontier; that is, their generation costs are high above the wholesale electricity prices (as well as the final prices of the kWh). This huge distance discourages investments and, thus, the diffusion of the new technologies and the improvements which are incorporated in those investments, even if there might be people who are enthusiastic about renewables for reasons beyond the purely economic ones.

In order to achieve the diffusion of the new-generation technologies, their profitability expectations need to be reinforced. At this point, through FITs or FIPs

[18]These innovations are so-called product innovation in the literature on industry life cycles [34, 78]. For reasons of simplicity, the innovations in the design of systems and subsystems have been included in this concept.

(together with other common incentives in the regulation of the electricity sector), the recovery of the investments in renewable energy plants is guaranteed (plus a profit margin) despite their comparatively higher costs. This reinforces the demand for equipment, which encourages the opening of manufacturing plants with a greater capacity of production per unit of time, which leads to economies of scale and experience. It also facilitates the incorporation of technical advances in the laboratory and the design offices. As mentioned above, the objective is to boost the downward trend of equipment prices and, thus, the cost of renewable energy generation.

However, the causal chain described in Fig. 4.11 contains many links in which the connection can break up. It should also be taken into account that there are many factors and collateral effects. Thus, an excessively expansive conjuncture may lead to the scarcity of some inputs, leading to an increase in its price and, thus, an increase in the cost of the energy generated. This occurred with polysilicon between 2004 and 2008. An excessive support also encourages speculation. It is not only about generating with renewables, but to gain money easily and fast with the sale and purchase of administrative authorizations and connection points. Investment booms regarding renewables in several European countries encouraged these practices. If the figure is analyzed from the perspective of RD&D, the results are reached slowly and, this, together with the uncertainty that accompanies all innovation efforts, may lead to the cancelation of programs [69: 53–88]. Thus, the promising research lines may be frustrated or the tuning of new manufacturing technologies may be delayed. It should not be forgotten that a technical achievement does not involve a commercially attractive design [53: 28]. There are many unforeseen combinations of factors which may delay the adoption of technological novelties [64, 63].

In this stage, the authorities of the leading countries favor technological diversity, given the availability of primary energy sources. RD&D centers and their programs and projects tend to cover many technological alternatives, irrespective of their progressive degree of maturity or distance with respect to the desired point of market launch [83: 34–35]. Notwithstanding, the most promising options are prioritized, especially if they exploit the most accessible resource(s) in the country. CSP is among them if there is a high DNI.

In order to understand RD&D policy for renewables, the following segmentation of the innovation process is illustrative[19]:

- Basic and applied research covers a wide range of this process: from the idea that some physical and chemical properties exploited, to the preliminary definition of a specific design, especially if the results from the laboratory are encouraging. Organic photovoltaic cells with graphene are currently in this stage.
- The development of the operative capacities of the system consists of the gradual improvement of the prototypes for a satisfactory operation, a guarantee of reliability, and reasonable expectations of costs. Parabolic trough plants with steam as a thermal fluid are an example.

[19]Own adaptation from Daim et al. [8], Grupp [20], Weiss and Bonvillian [83], and disregarding, for the sake of simplicity, the different feedbacks that exist between the different stages.

- The stage of demonstration is decisive from a technical point of view. This is the technology launch or introduction, although not the commercial deployment, of the new technologies. The installations, whose performance under real operation conditions is subjected to an intensive checking, are eventually connected to the electricity system. In this stage, the interest for the innovation in the manufacturing process accelerates.
- Precommercial diffusion refers to the connection to the electricity distribution grid of the first commercial plants. Its routine operation regime does not hide that its generation cost is not competitive yet. Demand-pull measures are useful here. The magnitude of the support in this precommercial stage depends on the urgency with which society perceives the convenience to deploy the new technologies. Although utilities can invest in renewable plants, firms from outside the traditional electricity sector normally lead the new-generation sector. CSP technology would have been placed in this stage until very recent times.
- Fully commercial. The ordinary regime for the exploitation of the plants, i.e., according to the conditions of the wholesale electricity market, is already profitable. There might be secondary improvements in this stage. The installations will be retired due to obsolescence or functional reasons. The diffusion leads to the replacement of existing plants by the new improved ones.

The low maturity of renewable energy technologies requires extending the public support beyond the basic and applied research stage, as we have tried to represent in the upper part of Fig. 4.12. The high risk of failure has been represented with a curve that does not follow the standard trajectory (thin line) but an upper one (thick line). The risk of failure in the stages of development and demonstration is so high that it discourages the massive arrival of private funds. In fact, until the end of the innovation cycle, the risk of failure is not reduced significantly. There are even serious doubts regarding its economic viability in the demonstration stage. Therefore, commercial diffusion requires demand-pull policies.[20] This is known since the 1940s in the realm of RD&D for defense and nuclear energy for civil uses.

The lower part of Fig. 4.12 shows that the provision of public funds is high in all the stages of the innovation process ([43]: Chaps. 6 and 7). This high and persistent provision of public funds for electricity generation technologies is justified in order to guarantee the security of supply and to mitigate climate change. Nuclear generation was prioritized according to the first argument, and the second argument was added later on.

The analysis of the features of demand-pull policies and instruments has to be added to the issue of the particularities of RD&D for renewable electricity. See Table 4.5. This would complete the view of the first stage of the process of the electricity transition.

The specific choice of an instrument by one country or region is related to ideological aspects, scenario analysis, the imitation of the experience of others, etc.

[20]The accumulated cost curve has a secondary role. Whatever is the public–private mix of support to a given RD&D project, its cost accumulates fast after the development stage.

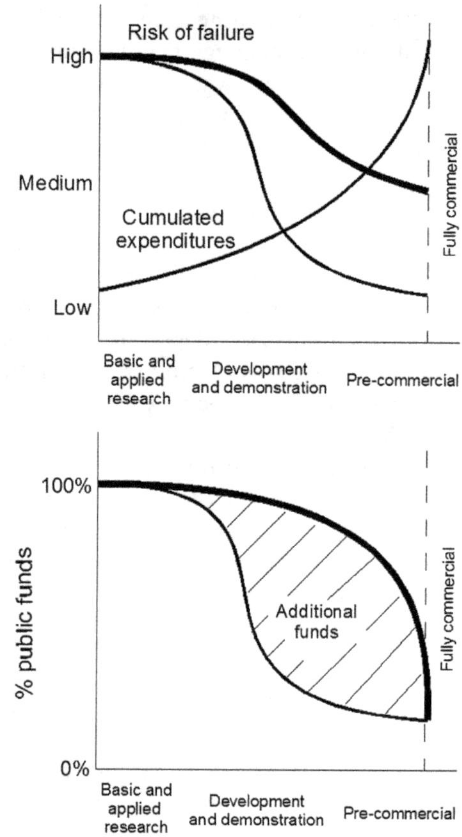

Fig. 4.12 RD&D stages and support for renewable energy technologies. *Source* Own elaboration

Table 4.5 Demand-pull policies and tools

Support mechanism	Main tool	Some features
Production-based (per kWh generated)	Feed-in tariff	Unforeseeable systemic rate
		Constant (for a given time)
		Decreasing at different rates
	Premium	Fixed
		Sliding
	Tradable green certificates	Different market conditions and production goals
Capacity-based (per kW installed)	Several reference plants and criteria for setting up the support amount according to a given variable (normally investment amount and/or operation costs)	

Source Own elaboration

Additionally, legal changes are also needed in order to facilitate the access of renewable energy plants to the grids and the access of its output to the wholesale market. Consumers are the ones finally paying for the policy, which is not a concern as long as the installed capacity is small.

The greatest challenge of demand-pull measures is to appropriately adjust the remuneration level to the reductions in the cost of equipments. This would avoid speculative booms, thus taking measures later which involve a sudden stop of the expectations of the sector. However, this is not an easy task. Not all the regulators are able to achieve a precise adjustment between both variables. Deviations may lead to ex-post cuts in the promised remuneration, which are interpreted as retroactive by those being affected by them. This may lead to lawsuits in the national courts and international organizations.

The impact of demand-pull policies can be easily modeled. The net cost of the promotion policy (V_T) is related to the capacity which is being accumulated over time, the dynamics of the tariff which is initially paid to investors in renewable energy plants, the annual updating of the tariff while the plant is active, and the evolution of the wholesale electricity price. With the aim to obtain a simple expression of such amount of costs, let us consider the following notation and simplifying assumptions:

- There is only a single renewable energy source, whose efficiency is constant.
- It is assumed that the amount of renewable electricity which is added every year is constant ($q_0 = q_t = \overline{q}$).
- The wholesale electricity price (w_t) goes down at a constant rate ρ (see below) so that $w_t = w_0(1+\rho)^t$, con $-\infty < \rho < \infty$.
- δ represents the annual rate of reduction in the preferential tariff. This reduction allows its adaptation to the reduction of generation costs (to simplify, it is assumed constant), whereas ϕ is the annual increase, for plants in operation, of the remuneration with which they were initially authorized. O&M grows over time.

Following the analysis carried out in Mir-Artigues and del Río [46: 434], for $t = T$, the promotion costs are equal to the accumulated amount of payments minus the observed reduction in the wholesale price. More specifically,

$$V_T = p_0\overline{q}\frac{(1+\phi)^{T+1}}{\phi+\delta} - w_0(1-\rho)^T$$

The behavior of this expression depends on the changes in the variables $\overline{q}$, ϕ, ρ, and δ. However, it can be demonstrated that the capacity which is added every year is the most relevant factor. In this case, if there is a boom, the resulting financial burden can be a slab for the electricity sector for years and, by extension, for society at large, unless cost-containment measures are taken, although they are never welcomed.

Although at the start of this stage the skepticism dominates and the option for renewables seems a laudable and expensive proactive effort, the fact is that some renewable energy sources, such as wind, PV, or CSP generation, have achieved a considerable degree of competitiveness. Even though the expectations of an increase in the price of traditional fuels have not been achieved, the improvements in the manufacturing processes of the equipments and the learning in their operation methods have been well above those imagined.

At present, it can be stated that many countries have deployed a volume of renewable energy which puts them at the end of this stage. Therefore, the description of the following two stages goes into unknown territory and the last one is a mere conjecture.

4.3.1.3 Second Stage of the Transition Period

The interval (t_2, t_3) in Fig. 4.10 represents the end of the structural change of the electricity sector. In this stage, a sharp reduction in the costs of renewable electricity is experienced and, thus, its presence in the electricity mix ends up being massive. This stage has two opposing trends:

- A maximum of the retail electricity price which is followed by its stagnation (maybe a reduction in some countries). The impact of the pioneering renewable energy plants, whose generation costs were high, reaches its maximum.[21] What to do, then, with the obsolete renewable plants whose lasting financial obligations distort the financing of the electricity system (the reason that the area *a* is extended beyond point *rgp*)? It is likely that controversial cost-containment measures will be implemented to reduce support costs. A better (but difficult to implement measure) is to replace these obsolete plants, which are not fully depreciated yet, with improved ones. Of course, this only affects the pioneering countries and, fortunately, the weight of the obsolete installed capacity is progressively lower.
- The growing presence of renewable electricity which enters the wholesale market at zero prices (since it has been remunerated with regulated prices) widely affects the merit order[22]: The supply curve shifts to the right, which further reinforces the pressure to close the generation plants with a higher cost. However, this effect changes depending on the hourly provision of renewable electricity, especially if it comes from variable renewable sources such as wind and PV. Furthermore, if dispatchable renewable energy sources have a small weight in the generation mix, and there is still not an important storage infrastructure, the variability of renewable electricity generation requires having a backup capacity which is able to face rampings and unexpected events. Maintaining a reserve of backup generation is expensive, and its retribution, which will probably fall on the consumers, is a thorny regulatory issue.

In this stage, the integration of traditional and new-generation sources is an issue of enormous complexity, which every country or region has to solve in a particular

[21] The gap between the wholesale and retail prices goes beyond T&D expenditures and other general costs of the electricity system.

[22] A critical explanation of the merit order effect can be found in Mir-Artigues and del Río [47: 143–144].

manner according to its starting mix, the available primary resources, and the weight of variable renewable sources. However, this singularity does not prevent us from drawing the contours of the issue. Let us consider the following notation:

D_p electricity demand peak
$\overline{D}$ average electricity demand
C capacity needs of the electric system
μ margin above demand peak in the conventional fuel generation system, usually 10%.

The required capacity is given by

$$C = D_p(1+\mu), \quad \text{or} \quad C = 3/2\overline{D}(1+\mu)$$

since, in a conventional generation system, it is usually assumed that $\overline{D} = 2/3\ D_p$. It should be added that the average load capacity factor is defined as $\overline{D}$ /C. Therefore, if, for example, $D_p = 100$ GW (or capacity required to generate the flow of electricity which is needed in order to cover peak demand), then $\overline{D} = 67$ GW and $C = 110$ GW. On the other hand, the average load capacity factor is 67/110, that is, 60%.

If it is assumed that this electricity system only has variable renewable sources (for instance, wind and PV), with its capacity factors (L) being $w_L = 0.3$ and $s_L = 0.15$, respectively, and with the renewable capacity (C_{R}) half wind and half PV, then the renewable average capacity factor ($\overline{L}$) would be $0.5{\cdot}0.3 + 0.5{\cdot}0.15 = 0.22$. If it is assumed that $C_{\mathrm{R}} = 50$ GW, then the renewable capacity which, on average, provides electricity to the system is 50 GW $\cdot$ 0.22 = 11 GW. This figure means that 11/67 GW, or a 16.4%, is the electricity which, on average, comes from the renewable energy sources being considered.

According to the previous reflections, the required capacity of the system with the presence of variable renewable sources should be

$$C_C + C_R \cdot \overline{L} = 3/2\overline{D}(1+\mu)$$

where C_{C} is the manageable conventional capacity ($C > C_{\mathrm{C}}$). In this way, the presence of electricity from renewable sources necessarily displaces the electricity from conventional plants. However, things are not so simple. Variable renewable energy sources have two important limitations:

- The coincidence (or not) between their generation peaks and the peaks in the daily (which, for example, in the case of PV is very good in warm and Mediterranean climates at noon, but very bad at sunset in winter in temperate and cold climates).
- The possibility that, in extreme cases, they are not available (in daytime hours, the absence of wind is quite common).

In order to solve such contingencies, several options exist:

- To increase the value of μ through conventional thermal plants which are able to face fast rampings, as it is the case with gas-fired plants.

- To build new (international) grids for transmission of the electricity, where they are non-existing.
- To increase the variable renewable provision by expanding the installed capacity over the widest possible geographical area (which maintains, or even reduces, its average capacity factor) in order to reduce the possibility of non-availability.
- To encourage dispatchable renewable plants, such as closed-cycle hydro, biomass, and CSP plants. They provide the ancillary services which are required by the electricity system in order to maintain its stability, while simultaneously reducing the need for conventional backup capacity.
- To take measures regarding interruptibility and demand response (see below).

As observed in Fig. 4.10, this stage ends when the renewable plants are competitive under the conditions set by the electricity market. This is the ultimate goal of renewable energy promotion schemes: to engage in an interaction dynamics between the average cost of a given j renewable technology $\overline{c}^j$ and its average preferential tariff $\overline{p}^j$ so that, after T years, $\overline{p}^j = 0$ that is, $\overline{c}^j \leq \overline{w}$ with $\overline{w}$ being the average wholesale electricity market price. Notwithstanding, it should be taken into account that those policies affect the later.[23] In other words, the aim is that the evolution of $\overline{c}^j$, $\overline{p}^j$, and $\overline{w}$ finally allows reaching the point *wsp* without serious distortions. This objective is not easy to achieve and includes a couple of key formal relationships between those variables. The next paragraphs discuss these relationships, leaving aside the factors which govern them and the vicissitudes over time.

To start with, let us consider the following notation and assumptions:

- The preferential tariff has a double dynamic: On the one hand, the tariffs of the already authorized projects increase and the initial tariffs for the new plants go down over time [47: 285–290]. If δ refers to the rhythm of reduction, whereas γ is the rate of updating of the average tariff,[24] the evolution of $\overline{p}^j$ per kWh of renewable electricity expressed in continuous time for t years is given by

$$\overline{p}_T^j = \overline{p}_0^j \frac{e^{\gamma t}}{e^{\delta t}} = \overline{p}_0^j e^{(\gamma-\delta)t} \quad 0 < \gamma < 1,\ \delta > 0$$

  This dynamic highlights both the willingness of the regulator to remunerate investments in renewables in a reasonable way, and to have the costs of the promotion policy under control.
- The term α denotes the rate of reduction of the average cost of the j-technology due to technical innovations, which improve the performance of the equipments and/or lower the manufacturing cost, to which learning and economies of scale also contribute. This all stems from RD&D efforts. The expression in continuous time corresponding to the cost dynamics is:

[23] The relationship $\overline{c}^j \leq \overline{w}$ only indicates that renewable energy plants may be profitable without any type of support. FITs or FIPs, and other support measures, are no longer required, as it has been represented in Fig. 4.10. The complexity that the variability of w entails has been ignored.

[24] There are plants with different ages, each with a specific remuneration regulation. Given the double dynamics, the average tariff for a given renewable energy will evolve according to the evolution of the initial tariffs and their updating rates.

$$\overline{c}_t^j = \overline{c}_0^j \mathrm{e}^{-\alpha t}$$

In order to strengthen renewable energy policies, the profitability of investments has to be as stable as possible over time. Let us assume, then, a constant value $r_t = r^*$. Or, in other words, let us assume that:

$$\frac{\overline{p}_t^j - \overline{c}_t^j}{\overline{c}_t^j} = r^*$$

which has to stand for the whole interval t ($t = 0, 1, 2, \ldots, T$) years. For a given technology, if $\alpha > \delta$, then the profitability of the projects increases, since the reduction in the costs offsets the reductions in the tariffs. Then, in case a reduction of δ is not foreseen, or that this reduction lags behind, investments can be very lucrative, which feeds bubbles. The opposite is the case if $\alpha < \delta$. Then, given that a horizon of $t = T$ years has been considered, the profitability of the investments in renewable will be constant throughout the period if the following condition is fulfilled:

$$\frac{\overline{p}_0^j \mathrm{e}^{(\gamma-\delta)t} - \overline{c}_0^j \mathrm{e}^{-\alpha t}}{\overline{c}_0^j \mathrm{e}^{-\alpha t}} = r^*$$

If this equation is solved for α and it is taken into account that $\overline{p}_0^j = \overline{c}_0^j(1 + r^*)$, then the final expression is derived, $\alpha = \delta - \gamma$. This means that the profitability (r^*) will stay constant as long as, for each technology, the rate of reduction in the cost and the difference between the regulated reduction in the price and the rate of updating of the tariff are equal.

On the other hand, given that the initial average wholesale electricity price $\overline{w}_0$ is too small for the investments in renewable energy plants to be profitable, a surcharge $\left(\lambda_0^j\right)$ is stablished, that is, $p_0^j = \lambda_0^j \overline{w}_0$, $\lambda_0^j > 1$. This surcharge is specific to each of the j-technologies. The advantage will disappear $\left(\lambda_T^j = 1\right)$ when the point *wsp* is reached, in $t = T$ years. It is reasonable to assume that, despite its volatility in the very short term and its changes of trend, the average wholesale market price experiences a reduction at the rate of ρ, due to the merit order effect. Thus, given that $\overline{w}_t = \overline{w}_0 \mathrm{e}^{-\rho t}$, in $t = T$ it is verified that $\overline{w}_t = p_T^j$.[25] Therefore, the point $t = T$ can be calculated, starting from the equality $p_T^j = \lambda_T^j \overline{w}_t$ and taking into account, in addition, that $p_t^j = p_0^j \mathrm{e}^{-\delta t}$. The result is the following expression:

$$t = \frac{\ln \lambda_0^j}{\delta - \rho}$$

[25] The different technologies do not reach point *wsp*. Those arriving there first will have a high probability to become dominant.

For example, for $p_0^j = 80$, $\overline{w} = 10$, $\rho = 1\%$, and $\delta = 10\%$, the value of λ^j is 8 and, therefore, $T = 23.1$ years.[26] This is the time lapse in which the trajectories of the preferential tariff and the wholesale electricity market price meet, given the initial distance which separated them. Obviously, it has been assumed that the evolution of the tariffs reflects the evolution of the renewable generation cost.

Three last remarks are worth making before closing this section. First, the dominance of renewable energy technologies will not avoid the presence of quasi-rents, or windfall profits,[27] in the wholesale electricity market. The electricity market quasi-rents are caused by the need to cover electricity demand with technologies whose generation costs are different. This is explained by the efficiency inherent to the different technologies, but also by the impact of exogenous factors (i.e., by a high price of fuels). Since the market price is set by the last plant which is needed to meet the demand at every moment (a price which only covers its variable costs), the rest of installations will benefit from differential rents. These rents will be higher the lower are those generation costs. Although those windfall profits will allow the accumulation of the resources required for the depreciation of the generation plants, quasi-rents will last commonly longer than necessary for capital recovery. Appropriate fiscal measures can be implemented in order to correct these rents [47: 119–121, 80].

The massive presence of renewable energy in the market, whose fuel costs are usually null (except in hybrid CSP plants and biomass plants), will not avoid the existence of quasi-rents, given that the levels of efficiency will be different (better locations and embedded technological advances). If the level of demand requires the activation of plants with comparatively high unitary costs, the rest of plants will keep on benefiting from differential rents. Moreover, it should be pointed out that, in the long term, the market price is set by the LCOE of the cheapest base load technology [38]. In appropriate places, CSP plants with TES are a strong candidate to play this role.

Secondly, once the point *wsp* gets close, many countries opt to organize renewable capacity auctions with the aim to reveal the lowest prices which investors are willing to accept. Regarding RD&D expenditures, there is no reason to suspect that they will be reduced in this phase. Apart from improvements in the manufacturing processes,

---

[26] In discrete time, $p_t^j = p_0^j(1-\delta)^t, 0 < \delta < 1$, and $\overline{w}_t = \overline{w}_0(1-\rho)^t$. After mathematical operations, we arrive at $t = \frac{\ln \lambda_0^j}{\ln(1-\rho)-\ln(1-\delta)}$. With the previous data, $T = 21.8$ years.

[27] Quasi-rents are a kind of differential rents currently associated with industrial activities: They happen when different technologies with different efficiencies are needed to satisfy the demand of a given product (i.e., quasi-rents are defined by the unit of output). In such a context, price is fixed by the less efficient plant and, thus, the other plants obtain increasing benefits with increasing efficiency levels. However, quasi-rents are temporary because, as time goes by, technological change modifies the efficiency order. For this reason, quasi-rents can also be understood as sustained windfall profits. It should be pointed out that there are many types of windfall profits. They normally occur due to unforeseen circumstances, such as an unexpected demand increase. For a detailed explanation on rents and quasi-rents, see Abraham-Frois and Berrebi [1: 113–118], Kurz and Salvadori [37: 277–320], and Salter [68: Chaps. 3 and 4].

there always be important technical aspects which can be improved, such as heat storage in CSP plants or the performance of photovoltaic cells.

Finally, demand-side generation can be generalized in this phase, which reduces the global electricity demand which is satisfied by traditional electricity and/or gas companies and, thus, their revenue expectations. Furthermore, the diffusion of self-generation leads to the concern of regulators and utilities about the remuneration of their investments in grids. In some cases, governments will be pressed to discourage on-site generation, although in other cases, some utilities may evolve to become energy service providers for the prosumers.

4.3.1.4 A Decentralized and Basically Renewable Electricity Sector

Achieving wholesale price parity confirms the success of renewables, but it is not the end of the story. Although the transitory period came to an end, deep changes in the technical and institutional configuration of the electricity sector are likely to happen. In Fig. 4.5, the fourth and definitive stage (t_3, t) has the absolute dominance of renewable electricity sources as its main feature. In this stage, the expectation is a deep reduction and later stabilization of electricity prices (in real terms). This evolution stems from the end of the preferential financing of the renewable plants and from the fact that the generation costs of the new installations are clearly competitive. Since there is dispatchable renewable capacity, all the electricity demand is progressively being covered only with renewables. The conventional backup capacity (zone *c*) is, then, reduced.

The key concept to be developed at this point is the distributed energy system (DES), in other words, an electricity system characterized by a scattered distribution of generation points (numerous small and medium-size plants, or distributed generation), to which distributed storage, electric vehicles, and devices which allow demand response at the industrial, commercial, and residential levels can be added. This is a complex technological framework which is connected to distribution grids under the supervision of refined and powerful information and communication technology (ICT) systems. These expectations are, however, totally uncertain. Therefore, the next paragraphs are mere conjectures, although the vision is shared by many experts.

In this period, the distribution grid becomes the key asset of the electric system. The large generation plants, which were a main element of the electricity system since the twentieth century, hand over the leading role to the smart distribution grids. The myriad of devices which make up the DES are connected to them. Although the transmission grid may still exist, the distribution grid can become a sort of federation of micro-grids, with even peer-to-peer platforms [57, 66, 82].[28] These micro-grids exchange energy with the rest of the electricity network. Two types of micro-grids may exist: those which gather generation and distribution in a given place under the control of a single owner, such as a university campus, a gated community,

[28]There might be cases of grid departure if the regulation allows it [5]. This possibility is currently unfeasible [33].

and a commercial or an industrial area, and those micro-grids which extend over a neighborhood or a whole city, or even beyond, pursuing the affiliation of more and more people and their generation and storage plants. The distribution grid, therefore, hosts autonomous sections and needs to be capable of managing energy flows in all directions, under the control of ICT systems in order to ensure its stability and reliability. Obviously, there will be some large conventional or renewable energy generation plants, all of them dispatchable (as CSP and biomass, perhaps coal with CCS and other technical possibilities which are unknown today), as well as the transmission grid in order to feed electricity to industrial areas and transport systems [49].

The ICT devices would facilitate that consumers have a greater control over their electricity consumption, responding to price signals. However, there are not likely to be many individuals and SMEs who are interested on the load management systems. There will even be fewer actors who might become potential suppliers of ancillary or capacity services to the electricity system. New firms which will carry out these functions will likely be created, and dominant firms in the ICT sector with branches dedicated to the intermediation between consumers and distribution and/or commercialization firms may enter this business. Those firms may install devices for the tracking and control of electricity consumption, which are able to send them a large quantity of information online which can be sold to the large generators and grid operators, as well as to regulatory agencies. In this sense, the electricity sector of the future could see a growing information asymmetry between utilities/ICT firms and regulators.

In this stage, it can be expected that saving and improving energy efficiency measures will be spread. In some countries or regions, it may mean that the global demand is stagnant or grows at a very small rate. This expectation also depends, as it is obvious, on economic and demographic growth.

The traditional distinction among industrial, commercial, and residential consumers, each with their own demand profile and passive behavior, could be blurred. The plausible abundance of prosumers may make it recommendable to establish charges for the use of the grids, according to the principle of cost causality [50]: The tariffs try to reflect the contribution of each user, whatever its size, to the costs of the grid (and its different components). This is a criterion which removes the problem of cross-subsidies. In order to do so, [60] propose that

- A reference network model (RNM) will have to be developed in order to ensure that the extension and reinforcement of the grids are planned in an appropriate manner. Without a detailed forecast, the grid could grow in a whimsical manner and thus the distribution costs could increase. This model allows the identification of the drivers of the distribution costs (new investments, amortizations, and O&M expenditures).
- The global cost (with the minimum profitability rate incorporated) is shared between the users depending on the time of use (per hour or fraction), assuming that this reflects their contribution to the total cost.

It is assumed, then, that the investments in grids are carried out in order to have a better knowledge about what is happening in them and to facilitate the fitting of the new users (generators and/or consumers). However, the change to a more decentralized electricity system does not imply that this will be more competitive. Although many DESs are owned by individuals, the larger generation plants, the grids, and the information flows will be in a few hands. There might even be alliances and mergers between utilities and ICT firms. Indeed, the traditional utilities, which have not played a main role in the development of renewable energy technologies or ICT systems, will hardly drive the transformation of the electricity system. It is perhaps more plausible to assume that those firms will focus on the management of the electricity grid under the supervision of the regulatory agencies (which will need to guarantee a level playing field for its access). Historical and institutional factors will determine the particularities of the electricity sector in this stage.

4.3.2 Values of CSP Electricity

The outline on the transition of the electricity sector represents a mere approximation. In practice, there are many details, some of which are specific to the country or region considered. In addition, the emerging elements mature over the years, although this process can accelerate at a given moment. The obsolete pieces slowly disappear. However, in spite of all the possible nuances, no one doubts that, when seen in a historical perspective, the configuration of the electricity system is changing. What we will see in 2050 will be very different to what we have seen a century before.

The aim of these pages is not to make predictions. The stylized model of the transition of the electricity system has been designed with a single purpose: to refine the analysis of the value of CSP generation, that is, to discuss its different effects on the transition of the electricity system and, by extension, on the economy and society. Thus, with this purpose in mind, we use the conceptual framework on the value of photovoltaic electricity provided by Mir-Artigues and del Río [47: 113–124] and apply it to the two hypothetical stages of the transition of the electricity system. This is a conceptual improvement since it is considered as a fact that the value of solar electricity will change with the transformation of the electricity sector. Although the attention falls on CSP generation, the comparison with PV is unavoidable.

To start with, let us consider Table 4.6, which lists the different components of the social value of CSP electricity (SV_{CSP}), together with the burden (−), benefits (+), or irrelevant effects (0) that they entail for society, the economic system, and the electricity sector. This is merely a theoretical exercise, although liable to empirical application since the evaluation takes into account the specific technological features of CSP generation and its degree of diffusion. All in all, the interpretation of the table has to consider that solar thermal electricity does not have a meaningful share in the

Table 4.6 Values of CSP generation and the energy transition model

<table>
<tr><th colspan="4" rowspan="2">Components of the value/effects</th><th colspan="2">Stages of energy transition</th></tr>
<tr><th>First</th><th>Second</th></tr>
<tr><td rowspan="7">Social value (SV_{CSP})</td><td colspan="3">Environmental value</td><td>+</td><td>+</td></tr>
<tr><td colspan="3">Welfare improvement value</td><td>0</td><td>+</td></tr>
<tr><td rowspan="5">Economic value (EcV)</td><td colspan="2">Cost of generation and early deployment</td><td>–</td><td>0</td></tr>
<tr><td colspan="2">Less hydrocarbon imports</td><td>+</td><td>+</td></tr>
<tr><td rowspan="3">Market integration value (MIV)</td><td>Merit order effect</td><td>+</td><td>0</td></tr>
<tr><td>Balancing services</td><td>0</td><td>+</td></tr>
<tr><td>Grid-related costs</td><td>–</td><td>–</td></tr>
</table>

power mix of any country, with a few possible exceptions.[29] As it was mentioned in the beginning of this book, the best locations for CSP generation are between 20° and 35° north latitude and south latitude, that is, the subtropical climate zones which are delimitated by the Tropic of Cancer and the Tropic of Capricorn, and where the larger deserts of the planet are located. This region, however, also includes countries with very different levels of economic development: from USA and the rich oil states of the Arabian peninsula to Mauritania. This has important implications with respect to CSP diffusion since knowledge and financial resources are unevenly distributed.

In the case of the first stage of the transition, the local and global scale effects of thermo-solar generation are very small or negligible whereas, in the second phase, their impact will depend on the extent of its diffusion in some countries.

The first term of Table 4.6 is the environmental value (EnV). The positive sign for both periods reflects the contribution of CSP generation to CO_2 emission reductions (see Chap. 2), both regarding electricity generation and producing of steam for industrial uses. The importance of the environmental value does increase not only when the installed capacity increases in the world, but also when more and more electricity from renewable energy sources is used in order to manufacture components and equipment for the plants.

Obviously, the inexistence of a clear penalty for the negative externalities caused by the emissions of greenhouse gases or the incapacity of the carbon market to set

[29]For example, in 2016, solar thermal generation in Spain (a country which currently has a relatively high share of CSP) was only ≥5% during 16.6% of the hours in that year, with only a few hours above 10% (the maximum was 10.15%). Given the high share of hours with very low (or even inexistent) generation, the annual average was only 1.96% (source: own elaboration based on TSO data from https://www.esios.ree.es/es/generacion-y-consumo). However, in the future, in countries with arid or desert regions the proportion of CSP electricity could be important.

a sufficiently high carbon price undermines the expectations of CSP generation as well as the other renewable energy sources. In this case, promotion policies have to be considered as second-best to the first-best carbon policies, which were deemed politically unfeasible.

CSP generation can have an important positive effect on the welfare of the residents of the region where the plant is located. Indeed, thermo-solar production can provide large quantities of desalinated water and, thus, contribute to the improvement of the surrounding agriculture and farming activities or to the urban supply of drinkable water. However, in contrast to photovoltaic generation, solar thermal electricity is not a modular technology, its operation is complex, and its maintenance is demanding. Thus, its role as a source of energy for the many rural or suburban communities in poor countries located in the tropics (whether to pump irrigation water or to generate electricity) has been very limited [61: 105–106]. A different issue is that CSP plants, which provide electricity and freshwater to its surroundings, entail an improvement of the living conditions of the residents which encourage migration toward those places.

The possibility of desalination justifies scoring a positive sign to the welfare improvement value of CSP. However, in the first phase of the transition process there is not any plant which produces desalinated water. If, as it seems likely, this use is diffused in the next years, the welfare value will turn from negligible to positive. It should be pointed out that this is only a possibility: The water for irrigation may be dedicated to crops for exports whose activities are carried out with a high degree of mechanization, or the supply may prioritize touristic areas and the richest districts in neighboring cities. If this is so, then the poorer population will not experience an improvement in its living conditions. Therefore, any specific diagnosis about the welfare impact of CSP with desalination will need to include the direct and indirect socioeconomic benefits of the drinkable water being produced.

There is no doubt that the main analytical concept of these pages is the economic value (EcV) of CSP, that is, its economic impact in terms of material and financial resources required to become a mature renewable technology and, as a result, its effects on the electricity markets. Its first component focuses on the Levelized-Cost-of-Electricity (LCOE).

The concept of LCOE is well known. It refers to the estimation of the generation cost of a plant (€/kWh or €/MWh), whether renewable or not, considering all the factors which affect its performance throughout its operative lifetime. However, the calculation of the LCOE is a delicate issue given the numerous elements involved, some of which are uncertain. This is the case with future fuel prices. Fortunately, CSP generation uses a free fuel (direct solar irradiation). Therefore, the CSP electricity LCOE may be defined in the following manner (adapted from [4: 70, 9: 3134]):

$$\mathrm{LCOE} = \frac{\sum_{t=0}^{t=T} \frac{C_t^*}{(1+i)^t}}{\sum_{t=0}^{t=T} \frac{q_{AC_t}}{(1+i)^t}}$$

As it can be observed, the LCOE is a ratio between the present value of the sum of the net costs of the plant (C_t^*) throughout its lifetime and the discounted flow of the energy generated. The costs include the initial installation expenditures, O&M costs, rental fees, charges and taxes, financial costs, and hybridization fuels. In case subsidies or any other incentives are incorporated in these variables, they should be deduced. In this definition, it is assumed that the electricity generated has the same value in all the hours of the year. Moreover, from the social perspective, the comparison of the LCOE of renewable energy and conventional technologies should consider the externalities [27: 15–16]. Finally, the costs of transporting the electricity to the consumption centers are not a part of the LCOE. This factor is considered in another section (see below).

The LCOE is calculated for specific plants. Therefore, even for the same technology, the LCOE of two plants will differ. On the one hand, the differences are smaller for conventional thermal plants than for renewable plants as the latter are very influenced by the climatic conditions. On the other hand, the LCOE significantly changes depending on the technical and economic assumptions used in its calculation. Therefore, we should pay attention to those assumptions, the origin of the data used, and the specific context in which they are interpreted.

It is also important to mention that the LCOE is an abstraction. It cannot be directly observable. It was a concept created in order to compare the generation costs of the different technologies from the point of view of investors, taking advantage of the fact that its output (electricity) is a physically homogenous good. Therefore, the LCOE can be interpreted as the minimum price that the owner of a plant should receive per kWh in order to cover the different costs of generation and still receive a normal profitability level [4: 70, 51: 104].

In Table 4.6, the value of the LCOE factor appears with a negative sign for the first part of the energy transition, whereas in the second part its impact is deemed null. The reason is quite simple: CSP generation is comparatively very expensive in the beginning, and thus, it needs strong support in the form of FITs or FIPs, whereas with technological improvements and learning, it is expected that its cost goes down and converges to the costs of the most competitive renewable and conventional technologies. When CSP plants do not receive any support (i.e., when they operate at market prices), the economic burden associated with early deployment will be null. It was negative when their deployment involved an extra cost for the electricity system. The trend of the LCOE for the coming years (see Chap. 3) points to a progressive reduction, until support is not needed. Of course, when this happens, it is likely that there will be plants in operation which have been deployed years before, which will continue to receive a preferential remuneration for the period envisaged in the regulation. Perhaps, measures to modernize these obsolete plants may be implemented.

The following component of the EcV is the lower fossil fuel and uranium imports which are allowed by an increasing deployment of CSP plants. The impact due to savings in imports is always positive, although its magnitude grows with CSP installed capacity. Taking into account its geographical conditions, CSP generation may be a key in order for countries with a strong solar irradiation to reach electricity self-sufficiency (to which the other solar sources, i.e., photovoltaic generation, will also contribute).

The entry of growing CSP volumes in the wholesale electricity market has different effects which are encompassed under the term market integration value (MIV). This component refers to the benefits and costs of thermo-solar electricity integration in the current managing of electricity market (assuming that this institution exists, as it is the case in countries with a liberalized electricity sector). The MIV includes three effects: merit order effect (MOE), balancing effect (BE) or service, and grid-related costs (GrC).[30]

Since the analysis of MIV is a complex issue, the following assumptions are adopted in the following pages:

- The country or region has very good direct solar irradiation.
- The limitations in forecasting the Direct Normal Irradiance have been completely solved. Different prediction methods have been studied, and as a result, DNI forecasting has hugely improved in these later years [36, 55, 71, 79].[31]
- There is a transmission grid which connects the production places to the consumption areas which are probably located far away.

In order to analyze the issue of the integration of variable sources, it should be clear from the start that capacity is not a proxy for flexibility. All electricity systems have some level of variability and uncertainty. Indeed, load changes over time (season, day of the week, and hour), sometimes in an unpredictable manner. Conventional resources can also fail without prior notice. However, variable renewable sources can vary in a way previously unknown for the SO, which have also difficulties to perfectly forecast them. As a result, there can be frequency deviations, load drops, energy curtailments, and price volatility, among other distress signals. These problems can be prevented if there is enough availability of ramping and fast response, transmission capacity (bottlenecks were removed), access to peaking plants (reserve capacity), load management, and so on. All of them are coupled with flexible system operations, that is, decisions that can be made closer to real time.

The value factor is an indicator which shows the interest that the electricity from a given source has for the electricity sector. Given the strict conditions under which the electricity system operates (equilibrium between generation and consumption, stable voltage, etc.), a maximum concern of SO is to have reliable sources, i.e., those which are capable of providing the needed electricity at a specific moment. From this point of view, there is not a perfect technology, although the cyclically intermittent and variable ones cause the greatest headaches. This is the case, for example, of solar photovoltaic generation. The additional costs that it may entail for the electricity system justify the statement that its LCOE is not a sufficient indicator of its value. The value factor, then, tries to quantify the cost of integration; i.e., it indicates the

[30] See [47] for a detailed definition of balancing deviations. See Hirth [23], IEA [26: 28–31, 34 and 67–82], and Mehos et al. [44: 6–9] for a complete analysis on the impact of variable generation on the management of the electricity system.

[31] The detailed prediction of the DNI requires collecting data for months. Its average annual values fluctuate up to ±15%. Having real time on ground and satellite data is also important.

difficulties in managing variable energy sources in the electricity system (see, e.g., Hirth [21, 22], IEA [26: 22–24], MIT [51: 104–106]). The value factor is the ratio of the average price per kWh received in the wholesale market by renewable generators divided by the hourly average market price during a certain time period (a year). In order to calculate the former, the revenues obtained by all the renewable plants in each hour are added and divided by the annual quantity of renewable electricity being sold. Therefore, this ratio shows the proportionality between the price received by the renewable generator and the market price, all over the yearly hours.

If we consider the case of PV generation, its value factor is greater than 1 when its degree of penetration is very small (<5%), given that the producers are in the best position to cover the peak of electricity demand in the middle of hot days (due to the high consumption of air-conditioning devices). However, when more and more solar PV electricity enters the system, problems in the management of surpluses in the central hours of the day emerge. There are also problems due to the lack of PV generation in the cold sunsets of temperate latitudes, when there is not any solar light and electricity consumption grows fast [3: 18–19, 47: 133–143].

The diagnosis regarding the integration of solar thermal electricity in the market is very different: The hybridization of CSP plants makes them a dispatchable source, which is a feature that is reinforced with TES [14: 10-35/10-36, 16, 44]. The hypothetical evolution of the value factor of CSP electricity is shown in Fig. 4.8. The thick discontinuous line reflects the value factor of CSP, whereas the thinner line refers to PV.

Figure 4.13 distinguishes between the first and the second stages of the electricity transition. The contribution of both solar technologies to the management of demand peaks in hot middays justifies that its initial value factor, that is, when the installed capacity is modest, is the same and above 1. As PV capacity increases, the management of the electricity system becomes more complicated. PV electricity generation is concentrated, especially in the hours with the highest irradiation. Its value factor goes down fast. In contrast, since the heat can only be stored for a limited number of hours, it can be assumed that the value factor of CSP could go down although such reduction is not so sharp. If regulation would allow total hybridization, the value factor of CSP in this first transition stage would not fall below 1.

The aim since the first CSP plants has been to saturate the capacity of the power block the maximum number of hours during the day. Thus, the size of the solar field, given the DNI, is adjusted in order to store the possible surpluses of heat and, from this, the period of only-solar operation can be extended. Therefore, in the second phase of the electricity transition, CSP plants are designed in a way that the volumes of TES allow operating in solar mode without interruption 24 h a day. Thus, it can be expected that the value factor of CSP generation increases to reach the level of dispatchable technologies (value factor equals 1). This is shown by the figure. Solar PV generation may also move in that direction, although the maturity of the electricity storage technology lags behind thermal storage. In the competition between both solar sources, one (PV) has an advantage in terms of location (almost the whole planet, since it can also operate with diffuse irradiation), and the other (CSP) has an advantage in terms of easiness to store energy, which allows it to be a

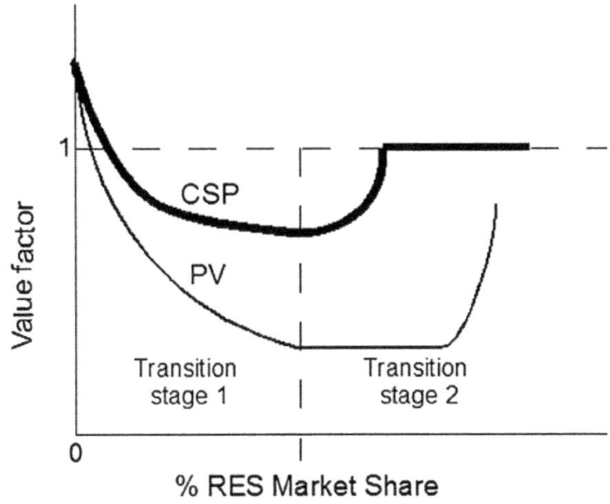

Fig. 4.13 Value factor of CSP electricity. *Source* Own elaboration

dispatchable generator. In reality, both complement each other: The dispatchability of CSP makes it feasible to have high levels of penetration of variable renewable, especially solar PV (see Denholm et al. [13: 38–40] and MIT [51: 198–199] for the case of USA). The combination of solar PV generation and CSP with TES allows solving demand peaks at middays in the summer through PV generation, whereas in the colder months and with fewer hours of sun, CSP allows overcoming the ramping when it is getting dark. This is shown in Fig. 4.14 where generation extends for 24 h, with a central day interval in which all the thermal contributions come from the solar field. The surpluses stored are used for electricity generation at night, which is never interrupted. Hybridization is not needed, except perhaps as a security measure in order to sustain the HTF temperature and in the event of breakdowns. Nevertheless, there are two time intervals at the start and the end of the daylight hours when the thermal flow of the solar field (which is below the one which is necessary in order to satisfy the rated power of the turbine) is combined with the one coming from the TES.

In this context, the variability featuring wind generation, which includes the interruption of its production (in the absence of wind or in the presence of strong winds), also increases the value of the dispatchable CSP electricity. However, the specific form of such complementarity between the electricity generation sources depends on the particular mix of each system. This is an issue which should be analyzed case by case with complex simulation models.

In case the generation capacity through the accumulated heat reaches the next day, CSP plants will displace conventional sources. This is a fact that can be generalized in the second phase of the electricity transition. If CSP generation does not slow down, the displacement of plants due to their comparatively higher costs could be permanent and definitive. In this case, the contribution of CSP to the merit order effect is out of doubt.

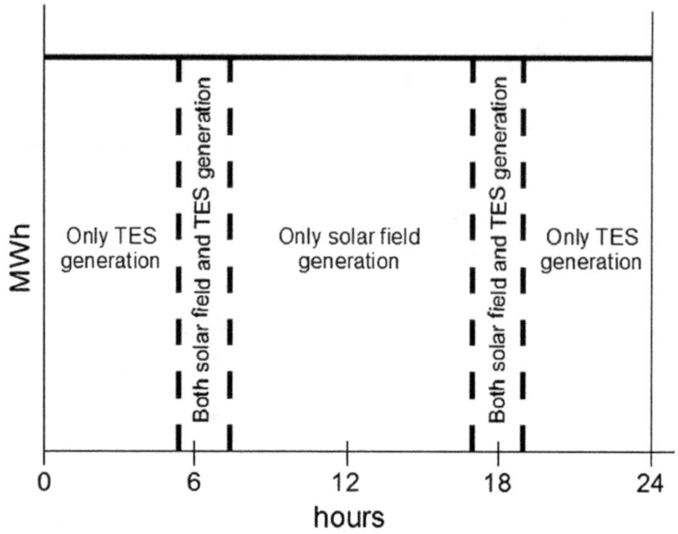

Fig. 4.14 CSP + TES 24/24 generation. *Source* Own elaboration

However, the merit order effect may end up being detrimental to CSP generation in the long term (a sort of cannibalization effect). The sustained reduction in the wholesale market prices would damage existing plants and would bring doubts about the profitability of new investments. The reduction in equipment prices (due to technical improvements and economies of scale) and, hence, electricity generation prices, would facilitate CSP diffusion, but would worsen the revenue expectations of planned plants as well. Therefore, the fitting between the successive reductions of the average prices in the wholesale electricity market and the improvements in the efficiency of the successive generations of renewable plants (until the possibilities of the technological paradigm are exhausted) should be addressed. In the best locations for thermo-solar generation, both trends should not put at risk the profitability of new plants, given electricity demand and its oscillations.

In general, and seen in perspective, the condition of entry into the market of the consecutive generations of renewable plants of a given *j*-technology can be written in a simplified manner as follows

$$w_0 \frac{q^j}{\Lambda} T \mathrm{e}^{-\rho t} = \frac{I_0^j}{\Lambda} \frac{r(1+r)^T}{(1+r)^T - 1} \mathrm{e}^{-\xi t}, \quad t = (1, 2, 3, \ldots, T)$$

The downward trend of wholesale prices, $w_t = w_0 \mathrm{e}^{-\rho t}$, leads to a downward trend of the revenues, assuming that the performance of new plants, that is, its annual production (q^j) per kW installed (Λ), is constant during the T years of its useful life. Since, for simplicity reasons, maintenance costs have been ignored, those revenues per installed kW have to allow the depreciation and profitability of the initial investment per kW (I_0^j/Λ), whose amount also goes down (at a rate of ξ). The new plants which comply with this condition will be able to access the market despite the reduction of wholesale prices.[32] This means that CSP, even in the best regions, needs to make a constant effort in order for the successive generations of plants to achieve higher levels of competitiveness. Obviously, the degree of maturity should not fall behind other renewables, especially PV.

Regarding balancing services, a plant without storage and/or hybridization has low probabilities to participate in those services.[33] If it has storage, then it can offer ancillary services, such as contingency/flexibility reserves, stabilizing frequency, and so on. This provision requires taking into account the ramping capacity and faster (less than hour) scheduling of solar thermal plants [72: 1].

The MIV analysis of CSP cannot conclude without addressing the issue of the impact of the cost of the transmission line of the electricity generated. The GrC might be the Achilles heel of CSP generation: It is expected that new investments in transmission lines will be needed to deliver the electricity produced by CSP plants. In fact, the best locations for CSP are unfortunately very far (hundreds of kilometers) from consumption areas, that is large cities and industrial centers. The exceptions are the countries in the Middle East, Maghreb, Sahel, Botswana, and Namibia and all those countries whose territory is mainly a desert, as well as some large urban areas located in far arid regions, such as Las Vegas, Iquique, or Yinchuan. In all these cases, the possible CSP plants can be deployed close to consumption areas. However, in the tropical countries, the population and economic activities prefer to be located in areas with moderate temperatures and safe access to water. The deserts are considered as very remote regions without substantial human activity (perhaps with the exception of mining activities, which employ only a few thousands of people). They are appropriate locations to install a thermo-solar plant, although a long transmission grid which reaches highly populated areas will need to be built. This is the case of Australia, Chile, China, South Africa, USA, etc.

The high voltage lines entail a very high investment, due to the cost of terminals as well as the fact that the wires need to be extended for hundreds of kilometers (especially if they are under water, where the savings in towers are more than offset by the strength which the wire requires). However, HVDC transmission lines over long distances are cheaper than the HVAC lines of the same distance, although the cost of the HVDC conversion equipment at the terminal stations is much higher, as shown by Fig. 4.15.

It is difficult to provide representative numbers of the cost in both cases, since each line represents a particular case. However, for the same path, the cost per kilometer of wire (whether on the air with supporting towers or under water) for an HVDC line is usually 1/3 of the cost of an HVAC line. The conversion equipment, on the

[32] As it is obvious, although it has been assumed for reasons of simplicity that only the investment per installed kW goes down, the increase in the efficiency of plants (more kWh per kW) is another factor that should be taken into account.

[33] The improvements in DNI predictions have encouraged the participation of CSP plants in the day-ahead electricity markets. Another measure to prevent high penalty payments for not achieving the predicted generation is to make CSP part of the portfolio of a market agent.

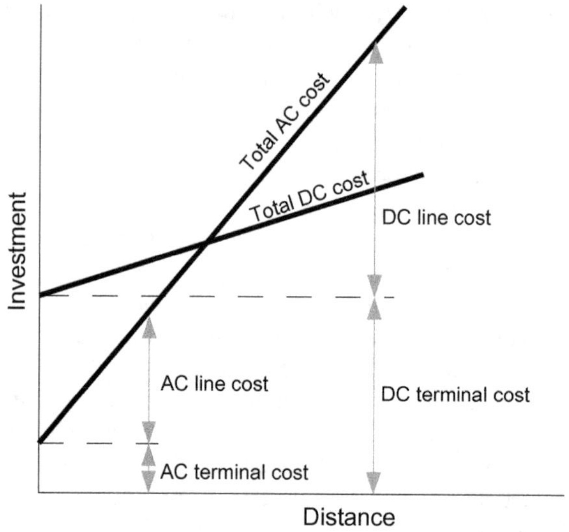

Fig. 4.15 Power transmission cost over long distances. *Source* Own elaboration

contrary, usually costs between 3 and 4 times more in the case of HVDC. The losses are also different: Regarding an aerial line, the losses are between 6 and 8% in the case of HVAC and half of those values for an HVDC. The losses for a line under water are very different: For example, for a submarine cable of 135 kV AC the loss is 18%, but for 400 kV DC it is 0.85%, both of 300-km length [30: 33–41]. Since the investment in transmission lines is on the order of billions of €, the distance between the generation and the consumption points the type of space to be crossed and the weight of losses are factors which need to be taken into account when choosing one or the other line. All in all, the final costs of the MWh of CSP can be affected by the cost of transmission or, rather, by the way in which its construction is financed and how such cost is distributed among the users of the line and the different consumers. This is a complex issue (see MIT [49: 88–96] and Rivier et al. [65: 293–309]).

References

1. Abraham-Frois G, Berrebi E (1980) Rentes, rareté, surprofits. Paris, Economica
2. Baharoon DA et al (2015) Historical development of concentrating solar power technologies to generate clean electricity efficiently—A review. Renew Sustain Energy Rev 41:996–1027
3. Baker E et al (2013) The economics of solar electricity. Energy Institute at Haas working paper series, p 240. Berkeley (Ca). Available at http://ei.haas.berkeley.edu. Accessed December 2013
4. Borenstein S (2012) The private and public economics of renewable electricity generation. J Econ Perspect 26(1):67–92
5. Bronski P et al (2015) The economics of load defection: How grid-connected solar-plus-battery systems will compete with traditional electric service, why it matters, and possible paths forward. Rocky Mountain Institute and Homer Energy, USA. Document available at www.rmi.org. Accessed July 2016

6. Caldés N et al (2009) Economic impact of solar thermal electricity deployment in Spain. Energy Policy 37(5):1628–36
7. Cocco D, Cau G (2015) Energy and economic analysis of concentrating solar power plants based on parabolic trough and linear Fresnel collectors. Proc Inst Mech Eng Part A: J Power Energy 22(6):677–688
8. Daim TU et al (eds) (2015) Policies and programs for sustainable energy innovations. Renewable energy and energy efficiency. Springer, Heidelberg
9. Darling SB et al (2011) Assumptions and the levelized cost of energy for photovoltaics. Energy Environ Sci 4:3133–3139
10. De la Tour A et al (2010) Economic analysis of the CSP industry. Paris, CERNA, Mines ParisTech and Agence Française de Développement. Available at http://www.cerna.mines-paristech.fr/Donnees/data07/704-CSP_report.pdf. Accessed Nov 2012
11. Del Río P, Mir-Artigues P (2014) A cautionary tale. Spain's solar PV investment bubble. Winnipeg (Can), International Institute for Sustainable Development/GSI. Available at http://www.iisd.org
12. Deloitte (2011) Impacto macroeconómico del sector solar termoeléctrico en España. Madrid, Protermosolar
13. Denholm P et al (2016) On the path to SunShot: emerging issues and challenges in integrating high levels of solar into the electrical generation and transmission system. NREL/TP-6A20-65800. Golden (Co) National Renewable Energy Laboratory. Available at http://www.nrel.gov. Accessed Feb 2017
14. Drury E et al (2012) Chapter 10. Solar energy technologies. In: Hand MM et al (eds) Renewable electricity futures study. Volume 2: Renewable electricity generation and storage technologies, pp. 10-1–10-60. Report NREL/TP-6A20-52409. Golden (CO) National Renewable Energy Laboratory (NREL). Available at http://www.nrel.gov/analysis/re_futures. Accessed July 2014
15. ESTELA (2018) Overview of the european solar thermal electricity industry directory. ESTELA
16. Gauché P et al (2017) System value and progress of CSP. Sol Energy 152:106–139
17. GEA (2012) Global energy assessment. Toward a Sustainable Future. Cambridge University Press and IIASA, Cambridge and Laxenburg
18. Grand E, Veyrenc T (2011) L'Europe de l'électricité et du gaz. Acteurs, marchés, régulations. Paris, Economica
19. Greenpeace et al (2009) Concentrating solar power global outlook 09. Greenpeace, Hamburg
20. Grupp H (1998) Foundations of the economics of innovation. Theory, measurement and practice. Edward Elgar, Cheltenham (UK)
21. Hirth L (2013) The market value of variable renewables. The effect of solar and wind power variability on their relative price. Energy Econ 38:218–236
22. Hirth L (2014) The market value of solar power: is photovoltaics cost-competitive? Paper presented at the 2013 solar integration workshop in London. Available at http://www.neon-energie.de. Accessed May 2015
23. Hirth L (2018) What caused the drop in european electricity prices? A factor decomposition analysis. Energy J 39(1):143–157
24. Houghton J (2015) Global warming. The complete briefing, 5th ed. Cambridge University Press, Cambridge
25. IEA (2010) Technology roadmap. Concentrating solar power. OECD/IEA, Paris. Available at http://www.iea.org. Accessed Jan 2011
26. IEA (2014) The power of transformation. Wind, sun and the economics of flexible power systems. OECD/IEA, Paris. Available at http://www.iea.org. Accessed Sep 2015
27. IRENA (2015) Renewable power generation costs in 2014. International Renewable Energy Agency. Available at http://www.irena.org/publications. Accessed March 2015
28. IRENA (2018) Renewable power generation costs in 2017. International Renewable Energy Agency, Abu Dhabi (UAE). Available at http://www.irena.org/publications. Accessed April 2018

29. Izquierdo S et al (2010) Analysis of CSP plants for the definition of energy policies: the influence on electricity cost of solar multiples, capacity factors and energy storage. Energy Policy 38:6215–6221
30. Jayasinghe K (2017) Power loss evaluation of submarine cables in 500 MW offshore wind farm. Department of Earth Sciences, Uppsala University
31. Jorgenson J et al (2013) Estimating the performance and economic value of multiple concentrating solar power technologies in a production cost model. NREL/TP-6A20-58645. Golden (Co), National Renewable Energy Laboratory. Available at http://www.nrel.gov. Accessed May 2016
32. Jorgenson J et al (2014) Estimating the value of utility-scale solar technologies in california under a 40% Renewable Portfolio Standrad. NREL/TP-6A20-61685. National Renewable Energy Laboratory, Golden (Co). Available at http://www.nrel.gov. Accessed May 2016
33. Khalilpour R, Vassallo A (2015) Leaving the grid: an ambition or a real choice? Energy Policy 82:207–221
34. Klepper S (1996) Entry, exit, growth, and innovation over the product life cycle. Am Econ Rev 86(3):562–583
35. Komendantova N, Patt A (2014) Employment under vertical and horizontal transfer of concentrated solar power technology to North African countries. Renew Sustain Energy Rev 40:1192–1201
36. Kraas B et al (2013) Economic merits of a state-of-the-art concentrating solar power forecasting system for participation in the Spanish electricity market. Sol Energy 93:244–255
37. Kurz HD, Salvadori N (1995) Theory of production. Cambridge, Cambridge University Press, A long-period analysis
38. Lamont AD (2008) Assessing the long-term system value of intermittent electric generation technologies. Energy Economics 30:1208–1231
39. Lilliestam J (2018) Whither CSP? Taking stock of a decade of concentrated solar power expansion and development. Deliverable 4.2, MUSTEC project. Zürich, ETH Zürich
40. Lilliestam J et al (2017) Empirically observed learning rates for concentrating solar power and their responses to regime change. Nat Energy 2:17094
41. Lilliestam J et al (2018) Policies to keep and expand the option of concentrating solar power for dispatchable renewable electricity. Energy Policy 116:193–197
42. Lovins AB et al (2002) Small is profitable. The hidden economic benefits of making electrical resources the right size. Rocky Mountain Institute, Boulder (Co). Available at https://www.rmi.org. Accessed May 2008
43. Mazzucato M (2013) The entrepreneurial state. Debunking public vs. private sector myths. Anthern Press, London
44. Mehos M et al (2016) On the path to SunShot: advancing concentrating solar power technology, performance, and dispatchability. NREL/TP-5500-65688. National Renewable Energy Laboratory, Golden (Co). Available at http://www.nrel.gov. Accessed Apr 2017
45. Mehos M et al (2015) An assessment of the net value of CSP systems integrated with thermal energy storage. Energy Procedia 69:2060–2071
46. Mir-Artigues P, del Río P (2014) Conbining tariffs, investment subsidies and soft loans in a renewable deployment policy. Energy Policy 69:430–442
47. Mir-Artigues P, del Río P (2016) The economics and policy of solar photovoltaics generation. Springer, Switzerland
48. Mir-Artigues P, González Calvet J (2007) Funds, flows and time. An alternatives approach to the microeconomic analysis of productive activities. Springer, Berlin
49. MIT (2011) The future of the electric grid. An interdisciplinary MIT study. Massachusetts Institute of Technology, Cambridge. Available at http://web.mit.edu/mitei. Accessed Feb 2013
50. MIT (2015a) The MIT study: utility of the future. White paper. The MIT Energy Initiative/IIT Comillas, Cambridge. Available at http://energy.mit.edu. Accessed May 2016
51. MIT (2015b) The future of solar energy. An interdisciplinary MIT study. Massachusetts Institute of Technology, Cambridge. Available at http://mitei.mit.edu. Accessed May 2015
52. Musgrove P (2010) Wind power. Cambridge University Press, Cambridge

53. Nemet (2007) Policy and innovation in low-carbon energy technologies. Dissertation for the degree of PhD. University of California, Berkeley. Available at https://mywebspace.wisc.edu/nemet/web/Thesis.html. Accessed Jan 2010
54. Nemet GF, Baker E (2009) Demand subsidies versus RD&D: comparing the un-certain impacts of policy on a pre-commercial low-carbon energy technology. Energy J 30(4):49–80
55. Nonnenmacher L et al (2016) Day-ahead resource forecasting for concentrated solar power integration. Renew Energy 86:866–876
56. Palenzuela P et al (2015) Concentrating solar power and desalination plants. Engineering and economics of coupling multi-effect distillation and solar plants. Springer, Berlin
57. Parag Y, Sovacool BK (2016) Electricity market design for the prosumer era. Nat Energy 32, art. 16032
58. Pariente-David S et al (2009) L'initiative de la Banque mondial en faveur d'une montée en puissance des centrales solaires à concentration (CSP). Utilisation du Fons des Technologies Propres en faveur d'une approche coopérative pou faire face aux défis du Changement climatique et aux défis énergétiques des pays du Moyen-Orient et d'Afrique du Nord (MENA). Réalités Industrielles, pp 42–49, Nov 2009
59. Pasinetti L (1993) Structural economic dynamics: a theory of economic consequences of human learning. Cambridge University Press, Cambridge
60. Pérez-Arriaga IJ, Bharatkumar A (2014) A framework for redesigning distribution network use-of-system charges under high penetration of distributed energy resources: new principles for new problems. CEEPR WP 2014-006. MIT, CEEPR
61. Perlin J (1999) Form space to earth. The story of solar electricity. Aatec Publications, Ann Arbor
62. Pollit MG, Anaya KL (2016) Can current electricity markets cope with high shares of renewables? A comparison approaches in Germany, the UK and the State of New York. Energy J 37(2):69–88
63. Popp D et al (2009) Energy, the environment, and technological change. (NBER working paper series 14832). NBER, Cambridge. Available at http://www.nber.org/papers/w14832. Accessed Mar 2011
64. Popp D (2010) Innovation and climate policy. (NBER working paper series 15673). NBER, Cambridge. Available at http://www.nber.org/papers/w15673. Accessed July 2012
65. Rivier et al (2015) Chapter 6. Electricity transmission. In: Pérez-Arriaga IJ (ed) Regulation of the power sector. Springer, London, pp 251–339
66. Rodríguez-Molina J et al (2014) Business models in the smart grid: challenges, opportunities and proposals for prosumer profitability. Energies 7:6142–6171
67. Rodríguez-Serrano I et al (2017) Assessing the three sustainability pillars through the Framework for Integrated Sustainability Assessment (FISA). Case study of a solar thermal electricity project in Mexico. J Clean Prod 149:1127–1143
68. Salter WEG (1966) Productivity and technical change, 2nd ed. Cambridge, Cambridge University Press
69. Scherer FM (1999) New perspectives on economic growth and technological innovation. Brookings Institution Press, Washington DC
70. Schill W-P et al (2017) Prosumage of solar electricity: pros, cons, and the system perspective. Econ Energy Environ Policy 6(1):7–31
71. Schroedter-Homscheidt M et al (2013) Aerosols for concentrating solar electricity production forecasts. Requirement quantification and ECMWF/MACC aerosol forecast assessment. Bull Am Meteorol Soc. Available at https://journals.ametsoc.org/doi/abs/10.1175/BAMS-D-11-00259.1. Accessed Jan 2018
72. Sioshansi R, Denholm P (2010) The value of concentrating solar power and thermal energy storage. TP-6A2-45833. Golden (Colorado), NREL. Available at http://www.nrel.gov. Accessed June 2012
73. Sovacool BK et al (2016) Fact and fiction in global energy policy. Fifteen Contentious Questions. Johns Hopkins University Press, Baltimore

74. Stoddard L et al (2006) Economic, energy, and environmental benefits of concentrating solar power in California. NREL/SR-550-39291. National Renewable Energy Laboratory, Golden. Available at http://www.nrel.gov. Accessed May 2012
75. Trieb F et al (2005) Concentrating solar power for the mediterranean region. Final report, German Aerospace Center-Institute of Technical Thermodynamics/Federal Ministry for the Environment, Nature Conservation and Nuclear Safety. Stuttgart. Available at www.dlr.de. Accessed Apr 2010
76. Trieb F et al (2009) Global potential of concentrating solar power. SolarPaces Conference, Berlin. Available at http://www.desertec.org. Accessed Apr 2010
77. Trieb F et al (2015) North Africa case study. German Aerospace Centre (DLR), Stuttgart
78. Utterback JM (1994) Mastering the dynamics of innovation: how companies can seize opportunities in the face of technological change. Harvard Business School Press, Boston
79. Vasallo MJ et al (2017) Economic MPC applied to generation scheduling in CSP plants. IFAC-PapersOnLine 50(1):115–120. Available at https://www.sciencedirect.com. Accessed Jan 2018
80. Verbruggen A (2008) Windfalls and other profits. Energy Policy 36:3249–3251
81. Wang J et al (2017) Status and future strategies for concentrating solar power in China. Ener Sci Eng 5:100–109
82. Weiller CM, Pollitt MG (2013) Platform markets and energy services. EPRG Working Paper 1334, Cambridge Working Paper in Economics 1361. University of Cambridge, Cambridge. Document available at www.eprg.group.cam.ac.uk. Accessed Dec 2016
83. Weiss C, Bonvillian WB (2012) Structuring an energy technology revolution. The MIT Press, Cambridge

Chapter 5
Support for Research, Development, and Demonstration

This chapter addresses the issue of why support for research, development, and demonstration (RD&D) in CSP is needed (Sect. 5.1), what are the instruments out there to promote it (Sect. 5.2), what level of innovative activity in CSP has been achieved (Sect. 5.3), and what types of innovation can be expected to be performed in the future (Sect. 5.4).

5.1 The Need for RD&D Support

As a less mature technology, which is in the early deployment stage, CSP has a significant potential for improvement. In general, improvements in the technology would lead to cost reductions or higher revenues. These improvements could come as a result of RD&D or through learning effects during diffusion. In addition, the increasing adoption of a technology leads to cost reductions as a result of dynamic economies of scale. Therefore, combining support for RD&D and support for deployment are clearly justified for technologies at this stage.

An initial decision is how to balance support for RD&D and for deployment in order to avoid the problems experienced in the past with solar PV, where there was a clear imbalance: too much deployment support instead of RD&D [1, 25], although public support at the commercialization stage is inherently greater than for previous stages of the innovation process (see Chap. 4). Note that cost reductions are a result of both deployment and RD&D. Therefore, both types of measures are inherently interrelated. A sine-qua-non condition to facilitate diffusion of the technology is to reduce its costs. In turn, this will allow further cost reductions.

P. Mir-Artigues et al., *The Economics and Policy of Concentrating Solar Power Generation*, Green Energy and Technology,
https://doi.org/10.1007/978-3-030-11938-6_5

5.2 Instruments for the Support of RD&D

There are several types of technology-policy instruments to support RD&D in low-carbon technologies in general and CSP in particular. Technology-push instruments, aiming at influencing the supply of new knowledge, can subsidize technology development directly, e.g., through public RD&D funding in research centers or universities, or indirectly, through fiscal measures aimed at encouraging RD&D investments by firms. RD&D can be encouraged with RD&D subsidies, tax credits, and rebates. Demonstration can be supported with the funding of demonstration projects.

Generally, technology-push policies include public RD&D spending (direct funding and grants), tax credits to invest in RD&D, capacity enhancement for knowledge exchange, support for education and training, financing demonstration or pilot projects, market engagement, incentive programs/public procurement, strategic development policies, technology exhibitions/fairs and network creation/building [14]. Table 5.1 provides examples of instruments targeted at different maturity levels of the technologies and different stages of the innovation process.

More specifically for renewable energy technologies, IRENA [19] classifies innovation policy options for them in seven functional categories, each pertaining to a potentially critical function for public policy. Within these categories, there exists a wide range of tools. The following Table 5.2 provides examples of these instruments for the different functional categories and the RD&D stages (i.e., not market development and commercial deployment). Note that "creating markets" is included as a seventh policy function, given the feedback from market creation to previous stages

Table 5.1 Illustrative examples of instruments to encourage low-carbon technologies

Stage	Supply-push instruments
Basic and applied RD&D	RD&D subsidies, tax credits and rebates, inducement prizes, networking, formation of partnerships
Demonstration	Government-supported demonstration programs/projects Grants, loans and risk guarantees Public venture capital agents
Pre-commercial	Government venture capital funds Incubators Tax breaks Networking Public venture capital agents
Niche market and supported commercial	Tax breaks Investment subsidies for up-front investments Public venture capital "agents" Networking Soft loans

Source Mir-Artigues and del Río [25]

Table 5.2 Examples of supply-push instruments according to policy functions and relevance for the RD&D stages

Policy function	Example of instruments	Relevance for which stage		
		Basic Science and RD&D	Applied RD&D	Demonstration
Building competence and human capital	Advanced degree programs: postdoctoral fellowships	X	X	
	Technical education, industry apprenticeships: "upskilling"			X
Creating and sharing new knowledge	Iterative product development	X	X	X
	Cooperative public–private RD&D		X	X
Knowledge diffusion and collaborative networks	Incubation of seed-stage entrepreneurship		X	X
	Interactive learning networks for value-chain growth			X
Governance and the regulatory environment	Robust intellectual property protection	X	X	X
Developing infrastructure	Grid-connected RET demonstration parks			X
Providing finance	Technology "Valley of Death" (seed) finance		X	X
Creating markets	Public procurement			X

Source IRENA [19, 20]

of the innovation process. The instrument "iterative product development" is based on the feedback between users and technology developers leading to innovation [26].

A look at the policy instruments used in most countries significantly investing in RD&D around the world shows that some of these instruments are already being applied in current policy practice [25].

Regarding RD&D support (also termed "supply-push"), there is a large potential for innovation and many possible paths to improve CSP technologies (see below). Broadly, this could be in the form of direct RD&D support for public institutions specialized in CSP and/or in materials and components which are required for CSP. Or it could lead to support for RD&D in industry through fiscal incentives or other instruments. An optimal path is to combine both, leaving public RD&D for the initial stages of the RD&D process, where the existence of a market failure is more likely (see Del Río [7]) and support for private RD&D for the commercial applications since firms are closer to the market and are aware of and can respond better to market needs. In addition, support for innovative demonstration plants within a public–private collaborative framework and encouraging networking and collaboration are crucial support instruments for this technology.

As to the historical CSP RD&D funding, the EU provides an interesting example, with a long history of support in this context. According to a recent study commissioned by the European Commission on the impacts of EU actions supporting the development of renewable technologies, Europe has funded solar thermal technologies for many years.[1] Since 1998, when the Fifth Framework Program (FP5) started, these technologies have received €400 million for 168 research projects, and another €38 million for 16 projects on solar thermal in combination with other technologies (see Table 5.3).

According to the figures provided in Table 5.3, most of the funding was allocated to CSP, which received €245 million in total (€12 million under Framework Program (FP) 5, €6 million under FP6, €152 million under FP7 and, until March 2018, €75

Table 5.3 European funding of solar thermal technologies

Framework program	Solar thermal		Solar thermal and other RES	
	EU funding (million €)	Number projects	EU funding (million €)	Number projects
FP5	56.48	47	1.91	4
FP6	21.92	22	7.75	2
FP7	207.95	53	22.77	9
HORIZON 2020	113.15	46	5.93	1
TOTAL EU funding	399.50	168	38.36	16

Source Cordis—(H2020 includes all projects awarded and registered in Cordis up to mid-March 2018)

[1] Under solar thermal technologies, the authors include: CSP, solar heating and cooling, solar process heating and cooling, PV thermal and multiple RES technologies (that is projects in which solar thermal is one of the multiple renewable technologies).

million under H2020) and 61 projects (13 projects under FP5, 4 projects under FP6, 23 projects under FP7, and 22 projects under H2020). As to the fields funded in CSP projects, many different technologies have been funded: solar tower concepts using innovative heat transfer fluids (gases, particles, etc.), Dish Stirling systems, combined solar thermionic–thermoelectric systems, hybrid CSP plants, small/medium CSP plants based on organic ranking cycle systems, etc. A number of other projects also focused on cross-cutting issues such as water use reduction, materials technologies, innovative thermal energy storage systems, and advanced systems for improved operations and maintenance [15].

Currently, RD&D support at the EU level for CSP is provided by the Horizon 2020 program, which is the largest EU research and innovation program ever, with nearly €80 billion of funding available between 2014 and 2020.

As to the top recipients of EU funding for solar thermal technologies by country since 2008, Spain, Germany, Italy, and France are at the top of the list. When it comes to the organizations, the top three recipients are: Deutsches Zentrum fuer Luft-und Raumfahrt e.V. (DLR), Centro de Investigaciones Energéticas, Medioambientales y Tecnológicas (CIEMAT), and Fraunhofer-Gesellschaft zur Foerderung der angewandten Forschung e.V. [15].

When considering the total renewable technology funding, overall, solar thermal projects received 11% of the €3603 million awarded to all renewable energy technologies through the FP5, FP6, FP7, and H2020 programs (until March 2018). The highest share was in FP7, when it accounted for up to 12.6% of the funding.

The European Solar Thermal Electricity Association (ESTELA) believes that, with a strong focus on technological demonstration projects, RD&D support is of great importance for the solar thermal electricity sector and a main driver to reduce cost, increase efficiency, and improve dispatchability and the environment profile [10]. In the past, European entities have funded important programs/projects (see Fig. 5.1).

However, European funds have only had limited usefulness, according to ESTELA. They helped to catalyze early projects such as Gemasolar, Andasol 1, and PS10, but the European Investment Bank has been the single biggest contributor. According to ESTELA's President: "Our sector is one that requires support at the commercial level, through things like project finance and soft loans, not at the laboratory level" [6]. Currently, schemes like the INNOFIN are addressing this need/challenge.[2] This approach contrasts sharply with the situation in the USA, where the government has supported commercial-scale CSP RD&D through the Department of Energy's SunShot initiative [6]. Between 1998 and 2015, the USA provided EUR 37 million per year on average to solar thermal research, of which nearly all went to CSP technologies. Together with Spain, it is the global leader in CSP capacity deployment and has been driving technology development in the sector over the last decade [15].

ALINNE [2] argues that support in the form of innovative demonstration plants within a public–private collaborative framework and funding of RD&D projects with

[2] See EIB [9].

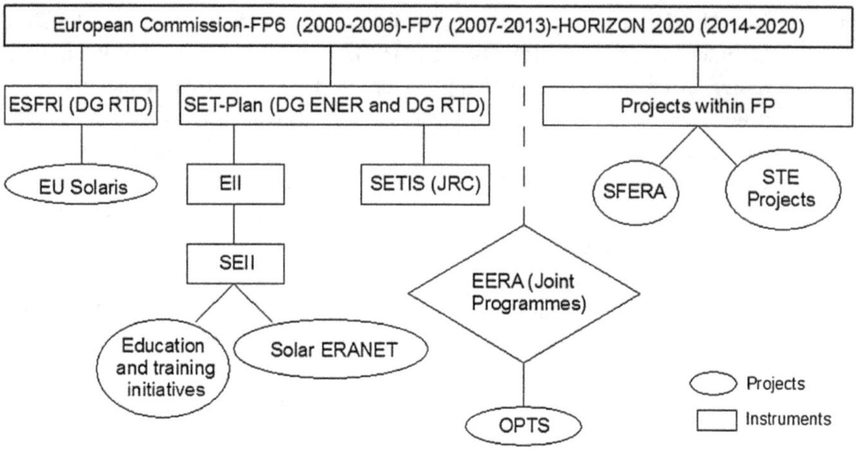

Fig. 5.1 Overview of the European schemes for solar thermal electricity funding. *Source* ESTELA (2012)

the participation of companies and (public) research centers have proven successful in the Spanish context and recommends their continuation in the future. In addition, there are other national and regional financing programs (see Del Río et al. [8] for further details).

A main supply-push instrument is supporting cooperation between public institutions and private actors and between private actors. Building networks of actors involved in RD&D has traditionally been a main instrument in RD&D policy [25].

Recent advances have focused on improving the coordination and cost-effectiveness of linkages between public and private actors, and across governance areas. For example, across the European Union, the European Research Area now administers a number of coordinated innovation investment activities that span public research institutes and private industry (e.g., the Framework Programs, European Research Council, the Competitiveness and Innovation Framework Program and the European Institute for Innovation and Technology). These activities aim to increase the interaction between research and commercialization activities [19, 20].

A main area in this context is international collaboration. One example of international CSP energy technology collaboration is the annual SolarPACES Conference, which is the largest European solar thermal electricity/CSP scientific conference and gathers industry, finance, and policy representatives. It facilitates technology transfer and knowledge exchange between the technology developing and adopting countries [4: 2451]. It has been a privileged place for exchanging information, sharing tasks, and above all (through the Plataforma Solar de Almeria run by CIEMAT) for sharing experience [16: 40]. The work program of SolarPACES includes six tasks (see Table 5.4).

Under the auspices of the European Energy Research Alliance (EERA), the CSP/Solar Thermal Electricity (STE) community was successful in establishing

Table 5.4 Tasks in SolarPACES

Task	Description (objective)
Task I. Solar thermal electric systems	It addresses the design, testing, demonstration, evaluation, and application of solar thermal electricity systems
Task II. Solar chemistry research	The objective is to develop and optimize solar-driven thermochemical processes and to demonstrate their technical and economic feasibility at an industrial scale
Task III. Solar technology and advanced applications	The objective is to advance the technical and economic viability of emerging solar thermal technologies and their validation with suitable tools by proper theoretical analyses and simulation codes and by experiments in special arrangements and adapted facilities
Task IV. Solar heat for industrial processes	The purpose is to provide the knowledge and technology necessary to foster the installation of solar thermal plants for industrial process heat
Task V. Solar resource assessment and forecasting	The task focuses primarily on the two most important topics in the field of solar radiation for solar energy applications: sound solar resource assessments and forecasting of solar radiation
Task VI. Solar energy and water processes and applications	The objective is to provide the most suitable and accurate information on the technical possibilities for effectively applying solar radiation to water processes, replacing the use of conventional energies

Source IEA [16: 40]

an integrated research program entitled "Scientific and Technological Alliance for Guaranteeing the European Excellence in Concentrating Solar Thermal Energy," also known as STAGE-STE. This EU seventh framework project promoted a specific activity focused on developing relations with public bodies relevant to CSP research, namely national research funding agencies, ministries, and other leading decision-making bodies. The purpose was to foster coordination and alignment of research programs at EU level, highlighting the added value of pooling resources for enhanced impact [5]. The program has received €10 million funding from the EU and a slightly higher amount from research performers and industrial players. There are 41 partners in the consortium, including nine non-European partners from those regions of the world with the highest CSP/STE deployment potential [11: 27].

Within STAGE-STE, national working groups (NWGs) were created to discuss and share experiences among each other. Such initiative proved to be an effective forum to promote information exchange and debate among CSP stakeholders and to connect entities with national and regional agencies. One common view among

NWGs is that there is a need for suitable and stable funding activities and for a specific research and technological development strategy for CSP at both national and European levels. According to Cardoso et al. [5], the EU agenda should foster the development of RD&D activities focused on lowering the technology and energy costs (reducing capital and operation and maintenance costs and increasing the produced energy) and improving its value (i.e., improving CSP technologies flexibility and ancillary services capability).

5.3 Data on Innovation Activity in CSP

Data on innovation activity can stem from different sources: patents, RD&D data, and expert statements. Each has pros and cons, which make them imperfect but yet very useful and widely applied measures of innovation. Patents are a useful indicator of the output of RD&D funding as they provide a direct measurement of the impact in terms of the novel knowledge generated. Furthermore, patent data are readily available in a standardized form, but also suffer from a number of drawbacks and raise issues that should be kept in mind when interpreting the results of the analyses which use this type of data [13]. First, the distribution of the value of patents is highly skewed to the right, since only a few inventions have a significant economic value. Second, not all inventions are patented and some firms might prefer a secrecy strategy to prevent imitation [4: 2446]. Filling a patent is not an objective of all research projects [15].

In contrast to the output feature of patents, RD&D data provide an "input" measure. RD&D informs about a crucial determinant of innovative activity (RD&D funds dedicated to innovation), but it does not say anything about the outcome of dedicating those funds to increase innovation, i.e., precisely about the results in terms of innovative activity.

Finally, expert assessment, despite being subjective, can provide very relevant insights about the possibilities of improvement in a given technology field.

5.3.1 Patents' Data

Unfortunately, and in contrast to other renewable energy technologies, studies specifically devoted to CSP patents are very scarce. A notable exception is Braun et al. [4]. The authors investigated the evolution of patent application counts in CSP compared to trends in overall patenting behavior. After a dynamic period, at the beginning of the 1980s, CSP patent applications experienced a downward trend, followed by a period of stagnancy lasting until 2000 and a subsequent slight increase in patenting activity. CSP technologies underperformed compared to the overall patenting activity at the European Patent Office (EPO): In 1978, they accounted for 0.54% of all patents and in 2004 for just 0.06%. Therefore, according to the authors, "the innovation performance of CSP is found to be surprisingly weak compared to the patent boom in other green technologies" [4: 2441]. The authors acknowledged that these results derived from a narrow definition of CSP technologies and that a different picture would emerge if CSP technologies were defined broadly, i.e., technologies which are important elements of CSP, but are not exclusive to CSP. According to this broad definition, patents followed the average patenting trend of the economy, moderating the pessimistic picture found for the narrow definition [4: 2449].

The authors also examined patent application counts at the country level, a common approach in the literature, to refine their analysis with a view on the leading inventor countries in this specific technology. The analysis showed that innovation leadership was highly concentrated in high-tech countries (the USA, Germany, and Japan), although only the USA had a large market potential for CSP applications. Thus, the authors concluded that CSP patenting was more influenced by research capacity and human capital than by high local DNI levels or high local deployment (or deployment support).

However, Braun et al. [4] are relatively outdated, in a technology field which has experienced a high dynamism in the last decade. More recently, Islam et al. [23] provide data on the yearly number of patents between 1981 and 2017 on CSP. The number of annual CSP technology-related patents was relatively constant between 1981 and 1993, in the range of 80–100 patents annually. Since 1993, the numbers increased exponentially to a record high of around 1600 patents being filed in 2013. Then, a sharp decline has been experienced, reaching around 900 patents by the end of the period (2017). The sharp increase until 2013 is interpreted by the authors as a tremendous progress in the technology.

Differently from Braun et al. [4], which carried out a search for CSP patents taking into account "search terms," a search taking into account the abstracts of the patent application has been carried out in order to include those patents which are registered in other classifications (i.e., chemical). The results of the search for patents in CSP are different depending on whether the WIPO or the PASTAT databases are consulted and also depending on the search terms included.[3]

Similar to Braun et al. [4], there is a considerable difference between the European and American data (in our case, European and international data).

The result of the search in the WIPO database carried out in August 2018 led to 6541 patents in "solar AND thermal AND (power or electricity)" and 1338 patents in "concentrated AND solar AND power."[4] Figure 5.2 provides the data in terms of number of patents per year in the last years (2008–2017). The data show an increasing trend in the number of patents over the years, with some stagnation in the middle of the period (i.e., between 2011 and 2016).

Regarding the countries which have filed most applications, China stands out, followed by the USA. The 2653 patents filed by China represent 40% of the total

[3]World Intellectual Property Office (WIPO) is a global forum for intellectual property services, policy, information, and cooperation [29]. PATSTAT is extracted from the European Patent Office (EPO) databases and contains bibliographical and legal status patent data from leading industrialized and developing countries [12]. We are grateful to Ch. Kiefer for support in this analysis.

[4]The search for the term CSP in the WIPO database leads to similar results regarding the evolution of total patents over time, their distribution per countries, and company applications.

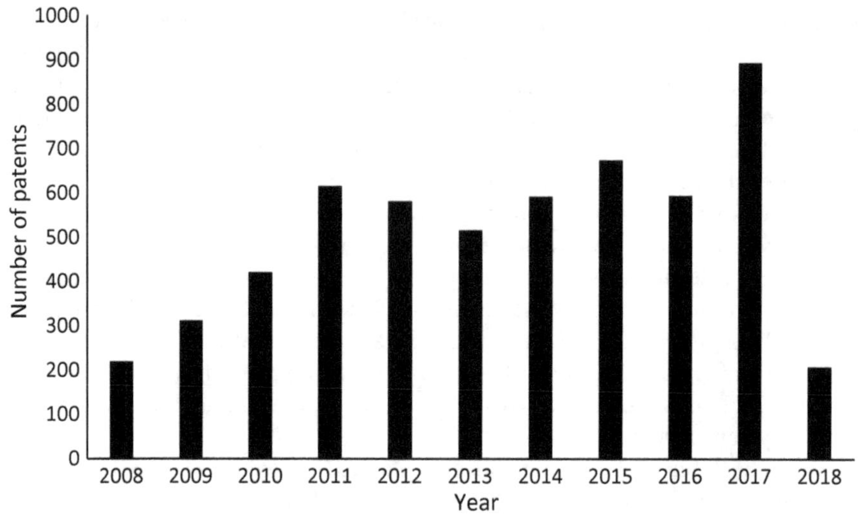

Fig. 5.2 Number of CSP patents in the WIPO database (2008–2018). *Source* WIPO database [30]. *Note* Data for 2018 only cover the patents filed until August 27, 2018

patents filed in the period. The USA (919 patents, 14% of the total), Japan (477, 7.3%), the European patent office (380, 5.8%), and Korea (230, 3.5%) follow behind. Chinese and German companies seem to be the leaders in this context, accounting for most patent applications during the period.

Regarding the PATSTAT database, the search has been carried out with two criteria—one is broader than the other. With the strict criterion (search for the PATSTAT patent queries with the terms "concentrated AND solar AND power"), 25 results have been found. Under the wider criterion, "solar AND thermal AND (power OR electricity)," 118 results have been found.

The analysis per countries shows that, when the narrow definition is considered, the USA (7 patent applications) and Germany (5 patents) stand out. When the criterion is broader, the most relevant countries in this context are Japan (32 patent applications), the USA (23), Switzerland (15), China (10), and Germany (8).

IRENA [22] provides data on the number of patents for solar thermal as well as for other technologies. According to these data (based on the INSPIRE web platform, [22]), as of 2016, there were 104 thousand patents in CSP technologies, slightly below wind energy and solar PV. This number for CSP represented a strong increase from 19 thousand patents in 2006 and 56 thousand patents in 2010.

An analysis of the patent filing with the IRENA INSPIRE tool/database (http://inspire.irena.org/Pages/patents/techprofiles.aspx) has been performed by the authors specifically for this book. Using the terms "solar thermal energy (Y02E 10/40)," the analysis has identified the top ten applicants in three selected periods ("last five years," "last ten years," "all the period"). The data show the dominance of Japanese and, to a lesser extent, Chinese firms, although the later have substantially increased their share in the last years.

Finally, Hoogland et al. [15] also performed an analysis of patents filed for solar thermal technologies in the EU. Their analysis showed that, during the 2000–2014 period, the number of EU patents filed increased up to 2010, and then it decreased to pre-2005 levels. According to the authors, such trend can be explained by the sharp increase of patents filed by China, rising from 500 patents/year in the early 2000s to more than 5000 patents/year from 2012 onwards. In addition, EU companies have started to question the benefits of patenting and are increasingly choosing to move fast to reap the benefits of their inventions rather than patenting.

When it comes to specific countries in the period from 2002 to 2014, their analysis shows that more than 8000 patents were filed in Germany, which made it the member state where most EU patents were filed (42% of the total number), followed by Spain (14%), France (11%), UK (5%), Austria (5%), and Italy (5%). Not surprisingly, these are the same member states that provided the largest funding. An exception is Italy, which provided most of the funding, but it is surpassed by five other MS in terms of patents.

5.3.2 Bibliometric Analysis

Bibliometrics (sometimes called Scientometrics) is the application of quantitative analysis and statistics to publications such as journal articles and their accompanying citation counts. Bibliometrics is used in research performance evaluation, in university and government laboratories, and also by policymakers, research directors and administrators, information specialists and librarians, and researchers themselves [27: 3]. Publications of research papers are a useful indicator to measure the output of RD&D funding, as there is enough data available to make a comparison between countries or regions in the world. "Moreover, publications have a close relation with public RD&D funding, allowing to differentiate the effect of public RD&D funding from private funding" [15: 21].

With a search term "concentrating solar power" in the SCOPUS database, Islam et al. [23] found a total of 15,998 patent documents. Their text-mining-based bibliometric analysis on CSP research shows that the total number of publications has continuously grown between 1981 and 2018. This growth has been significant in the last decade (2007–2017), as the research published after 2007 consists of 91.5% of all the papers published in CSP-related studies. In 2017, the highest number of papers (157) was published. In total, Islam et al. [23] have identified 114 major publication sources (including both journals and conferences). It is found that CSP-related research has mainly progressed through journals (except AIP conference proceedings) and that the top 15 sources are responsible for around 55% of all the papers, while the remaining 99 sources are responsible for 36% of the publications. The rest (9%) is from other 104 sources. *Solar Energy* and *Energy Procedia* are the two journals where the majority of the articles have been published in recent times, while *Renewable and Sustainable Energy Reviews* is the journal that has contributed progressively from 2009 [23: 1008].

According to Hoogland et al. [15], EU-based authors were involved in a third of the global publications between 1995 and 2017, making them the global leaders. Outside the EU, the largest number of publications had authors from China, the USA, and Canada. Combined, these authors were also involved in a third of the global publications between 1995 and 2017. China accounts for 20–25% of publications in recent years. According to the same study, within the EU, authors from Germany accounted for most publications (202) followed by Spain (200), France (157), UK (153), and Italy (139).

5.3.3 *Public RD&D Data*

Public support for RD&D has been crucial for the development of CSP, as it has also been the case with other renewable energy technologies, such as wind and solar PV [4]. Public RD&D data are provided by the publicly available IEA RD&D database [17]. The search for CSP data (code 313 "Solar thermal power and high-temperature applications") reveals that, in constant prices, the total amount of public support for RD&D in CSP accumulated over the period 1974–2016 has been million €4185. Several conclusions can be inferred from the data. First, such total amount of public RD&D can be considered very small compared to deployment support. Although total support for deployment in CSP over the years is not available, such figure is comparable to deployment support for CSP in a single year in a single country (Spain million €426 in 2011). Second, Table 5.5 shows a large concentration of public RD&D support in a single country (the USA, about 2/3 of total support over

Table 5.5 Accumulated RD&D budgets for CSP in the 1974–2016 period in seven IEA countries (million €, 2016 prices and exchange rates)

Country	Total (million €)	% over total
Australia	110	2.6
Germany	322	7.7
Italy	312	7.5
Japan	163	3.9
Spain	282	6.7
Switzerland	212	5.1
USA	2610	62.4
Subtotal	4011	95.8
Total IEA	4185	100.0

Source Own elaboration from IEA [17]. *Note* Only the seven countries with the greatest RD&D budgets have been included in the table. Data are not available for several years during the period. We have allocated a "zero" value in this case

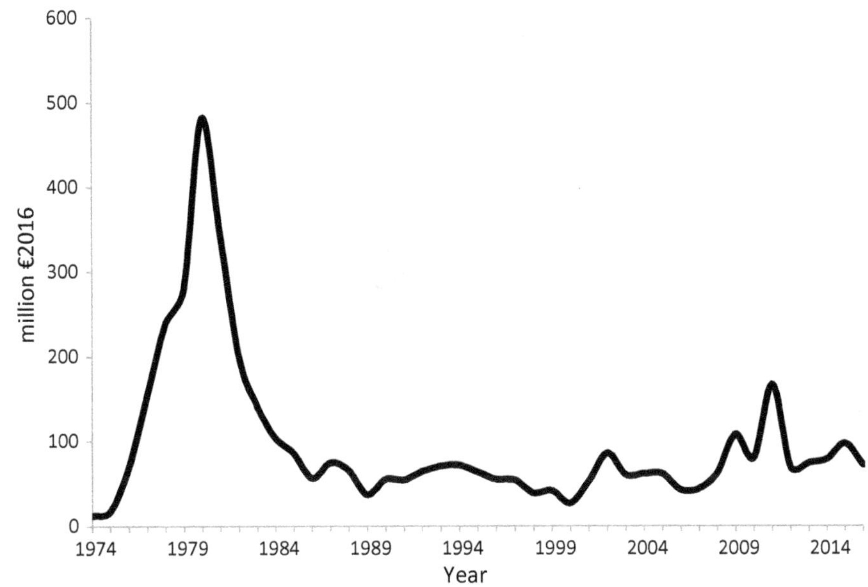

Fig. 5.3 Evolution of RD&D budgets in CSP in the 1974–2016 period (million €, 2016 prices and exchange rates). *Source* Own elaboration from IEA [17]

the period), with the second country (Germany) falling far behind. Third, the highest amounts of RD&D support took place in the late seventies and early eighties, and fell sharply afterward, recovering in the early 2000s (see Fig. 5.3), a trend observed also for other renewable energy technologies. However, this recovery has been modest, with ups and downs during the last subperiod, and in any case, absolute volumes have not reached the levels achieved four decades ago. Fourth, the pattern of public support for RD&D is similar across countries (see Fig. 5.4).

A disaggregation of total CSP RD&D budgets into RD&D and demonstration in the 1974–2016 period shows that an overwhelming majority of the funds (96%) has been dedicated to the first stages of the innovation process (research and development), whereas support for demonstration has been much less important in this regard (see Table 5.6), although it has gained some relevance in some countries by the end of the period. Australia, Denmark, and the USA concentrate around ¾ of demonstration funds in the period. Demonstration is likely to be more relevant in the near future in order to improve the technology and reduce its costs.

Braun et al. [4] provide data on the relative importance of CSP versus other solar public research funding. Arguably, a measure of the commitment to develop CSP is the ratio of public RD&D support for CSP to RD&D support for the other solar technologies, PV, and solar heating and cooling. They show that a downward trend for most countries can be observed. Since 2000, however, interest has revived, which the authors attribute to awareness on GHG mitigation due to the Kyoto protocol.

Finally, according to a recent study commissioned by the European Commission [28], when considering the resources devoted to RD&D for solar thermal during the

Table 5.6 Disaggregation of total CSP RD&D budget into RD&D support in the 1974–2016 period per country (million $, nominal)

Country	Research and development	Demonstration
Australia	112.007	32.907
Austria	10.738	0.5
Belgium	6.148	0
Canada	14.393	3.239
Czeck Republic	13.906	0
Denmark	73.367	23.788
Estonia	0	0
Finland	1.108	0
France	30.137	3.514
Germany	250.392	0
Greece	1.533	0
Hungary	0	0
Ireland	0.607	0.039
Italy	233.203	0
Japan	18.500	0
Korea	1.703	0
Luxemburg	0.507	0.147
Netherlands	22.481	0
New Zealand	0.868	0
Norway	121.500	0
Poland	0	0
Portugal	0.798	0.05
Slovakia	0.697	0
Spain	165.201	0
Sweden	17.700	13.1
Switzerland	205.326	7.051
Turkey	0.115	0
UK	0.160	1
USA	1506.466	34
EU	0	0
Total	2809.500	6

Source IEA [17]. Code "313 Solar thermal power and high-temperature applications"

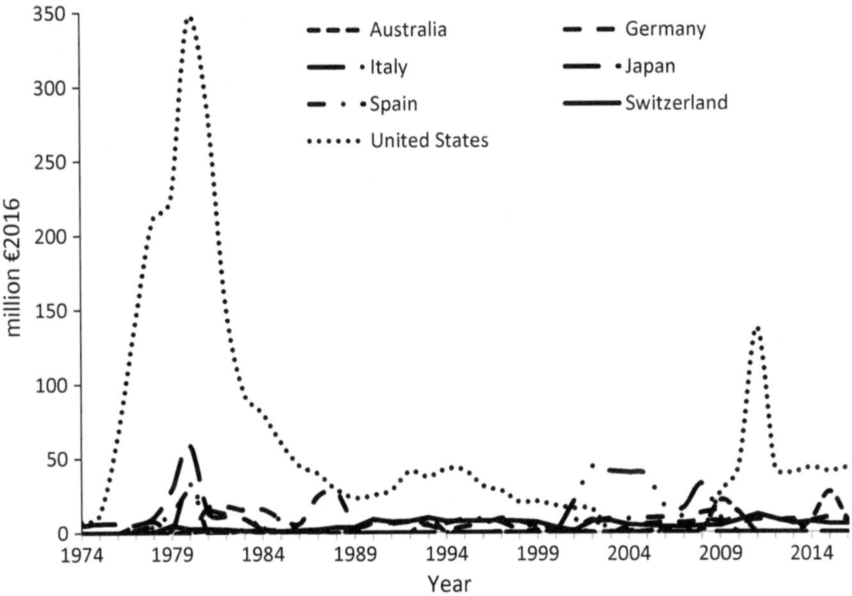

Fig. 5.4 Evolution of RD&D budgets in seven IEA countries in CSP in the 1974–2016 period (million €, 2016 prices and exchange rates). *Source* Own elaboration from IEA [17]

period 1996–2015 in Europe, the authors found that the largest member state funders were Italy, Germany, Spain, France, and the UK. Together, they accounted for 90% of all member states' RD&D funding. While some of these countries show high correlation with their solar resource (Spain and Italy), other less sunny countries were also very active in CSP. For example, the German Aerospace Centre (DLR) has been part of the Plataforma Solar de Almería since 1987 and the German Research Institute Fraunhofer ISE is another important player in CSP. Interestingly, Italy provided more funding to RD&D on CSP technologies than Spain, but did not install any CSP plants. This shows that member state funding of RD&D for solar thermal technologies is not the only driver for sector development. For the case of Spain, the main driver was the national deployment policy (see Chaps. 3 and 6).

5.4 Future Technological Improvements

Several contributions identify future technological improvements, generally based on the opinion of experts on the technology. For example, in a recent comprehensive review of the state of the art of CSP technologies, their current status and the research trends, Islam et al. [23: 1013] argue that there are several evolving research topics on

CSP, including the direct steam generation technology of the parabolic trough collector (for hybridization with other CSP technologies for electricity and thermal heat production), optimization of solar fields (solar multiple), estimation of Levelized-Cost-of-Electricity with sensitivity analysis, application of system advisory models in CSP plant development, implementation of organic Rankine cycle engine for heat and power production, investigation on supercritical CO_2 power cycle in CSP plants and performance analysis of calcium-looping and thermochemical energy storage in the CSP systems.

Zarza et al. [31] examine how technology innovation may reduce the LCOE from European CSP plants over the next 12–15 years. Its input data are closely based on the KIC InnoEnergy technology strategy and roadmap work stream published in October 2014, which led to a comprehensive set of discrete innovations and groups of innovations together with their potential impact on known reference plants, based on expert vision and knowledge. This study has a cost model in which elements of baseline solar thermal electricity plants are impacted by a range of technology innovations. The report identifies 64 possible innovations to reduce the cost of CSP plants. These are defined in terms of the applied technology type: parabolic trough collector, central receiver, and linear Fresnel reflector [31: 5]. For all the technology types, technology innovations in CSP would reduce the LCOE by 23.6% or more [31: 6]. Table 5.7 provides a list of the most relevant innovations which are expected to reduce the costs of CSP plants in 2025 for each technology type.

IEA [16] identifies several possibilities for further innovation in CSP technologies. These include incremental improvements in existing technologies as well as improvements in new, more radical technologies. Some innovations are considered key for cost reductions, including novel optic designs, new mirror materials and receiver designs, laminated reflective components glued on aluminum sheets (for parabolic troughs) and choices relating to the type of receivers (cavity or external), the number and size of heliostats, the number of towers associated with each turbine, and the size and shape of solar fields (for solar towers). More radical technologies include new thermodynamic cycles (supercritical steam cycles, Brayton cycles with a gas turbine or supercritical CO_2) [16: 23]. Higher working temperatures are a particularly relevant innovation avenue. This is key in order to increase efficiency in converting the heat into electricity and reduce storage costs. Improved efficiency also lowers the cooling load and the performance penalty caused by dry cooling. However, there are some trade-offs at stake, since higher temperatures increase the thermal losses of the receiver through convection and radiation and may require more expensive materials [16: 26]. According to the IEA's technology roadmap for solar thermal electricity [16], the following areas of improvement are possible (see Table 5.8).

Table 5.7 Most important innovations expected to reduce the costs of CSP plants in 2025, compared to a CSP plant in 2014

Parabolic trough • Improved and cheaper manufacturing methods and automated production of components • Advanced high-temperature working fluids • Improved solar concentrator design • High-temperature receivers • Tools for onsite checks of solar equipment • Efficient plant monitoring and control with continuous onsite checks of solar equipment • Software development at system level + 14 other innovations
Solar tower • Improved and cheaper manufacturing methods and automated production of components • Improved solar concentrator design • Software development at system level • Efficient plant monitoring and control with continuous onsite checks of solar equipment • Tools for onsite checks of solar equipment • Design and coating of central receivers • Software development at component level + 17 other innovations
Linear Fresnel • Improved and cheaper manufacturing methods and automated production of components • Tools for onsite checks of solar equipment • Efficient plant monitoring and control with continuous onsite checks of solar equipment • Advanced high-temperature working fluids • High-temperature receivers • Software development at system level • Advanced power cycles + 12 other innovations

Source Zarza et al. [31: 7]

For IRENA [18], key areas for cost reductions include the solar field (mass production, cheaper components, and improvements in design), the heat transfer fluid (new heat transfer fluids and those capable of higher temperature, direct steam generation), the storage system (closely tied to the heat transfer fluid, as higher temperatures reduce storage costs), the power block and the balance of costs (IRENA [18: 21]).

IRENA [21] considers key technological improvements which may lead to CSP cost reductions in a 2025 timeframe. For parabolic trough, two of these improvements are worth mentioning: (1) a switch from thermal oil VP-145 for the HTF of plants to solar salt (this enables higher process temperatures and reduces installed costs and LCOEs significantly); (2) greater aperture widths from 7.5 to 10 m by 2025 for the trough collectors [21: 91]. This leads to a smaller number of collectors for the same total aperture area, reducing capital costs. For solar tower technology, the innovations expected are less revolutionary compared to the HTF change for trough systems, since the use of solar salt for solar towers is already state of the art. Out to

Table 5.8 Technology development: Actions and milestones

Actions	Timeframes
1. Demonstrate molten salts as HTF in linear systems (parabolic trough and linear Fresnel) at large scale	Complete by 2018
2. Develop lightweight, low-cost reflector optics	Complete by 2018
3. Optimize heliostat size, solar field design, central receiver design, number of towers per turbine for 6–18 h of storage	Complete by 2018
4. Introduce supercritical steam turbines in CSP plants	Complete by 2025
5. Increase the energy in receiver tubes with innovative non-imaging optics for linear systems	Complete by 2020
6. Introduce innovative HTF: air, gas, nano-fluids in linear systems, fluoride liquid salts, air and particles in towers	Complete by 2025
7. Introduce closed-loop multi-reheat Brayton turbines	Complete by 2025
8. Develop and introduce supercritical CO_2 cycles	Complete by 2030
9. Develop hybrid PV-CSP via spectrum-splitting or PV topping	Complete by 2030
10. Intensify RD&D on solar fuels (gaseous, liquid, or solid)	2015–2050

Source IEA [16: 26]

2025, however, heliostat reflectivity and receiver efficiency are expected to improve, resulting in slightly higher operating temperatures (565–600 °C) and higher power block and overall plant efficiency levels [21: 93]. The overall efficiency of solar tower technologies can be expected to increase from 15.5% in 2015 to 18.3% in 2025, driven by improved availability and higher temperature levels in combination with supercritical steam cycles (such as the ones currently used in modern coal plants). However, for solar towers, the largest single driver for LCOE reductions is related to gains in the engineering, procurement, and construction experience [21: 102].

In its Strategic research agenda 2020–2025, ESTELA mentions three objectives defined by the solar thermal electricity industry (increase efficiency and reduce costs, improve dispatchability, and improve environmental profile). Three transversal research topics to reach those objectives include activities in three realms: mirrors (light reflective surfaces and anti-soiling coatings), heat (transfer fluids, low melting temperature mixtures, pressurized gases, direct steam generation with high-pressure absorber tubes and high working temperatures), and others (selective coating receivers with better optical properties, new storage concepts and better control, prediction and operation tools).

The European Strategic Energy Technology (SET) Plan includes a CSP RD&D agenda alignment and implementation of joint RD&D activities in SET Plan countries (EU Member States, Norway, Iceland, Switzerland, and Turkey) (see Table 5.9).

Table 5.9 RD&D activities included in the SET Plan implementation plan

• Improved central receiver molten salt technology
• Parabolic trough with silicon oil
• Next generation of central receiver power plants
• Advanced linear Fresnel technology
• Parabolic trough with molten salt
• Open volumetric air receiver
• Multi-tower beam down system
• Advanced TES
• Supercritical steam turbine
• Improved flexibility in CSP applications
• High-temperature Brayton
• Pressurized air receiver with storage

Source Blanco [3]

The overall goal of this plan is to facilitate the achievement of the EU climate and energy goals and to strengthen industrial competitiveness. This is being done by better coordinating national RD&D agendas on low-carbon energy and mobilizing the associated resources required [11: 19]. The goal of the CSP/STE implementation plan is to achieve significant cost reductions for existing technologies in the short term and to work toward developing the next generation of technologies in the longer term. The plan has been drafted in a working group composed of interested SET Plan countries, stakeholders, and the European Commission [11: 80]. The plan identifies five strands of action: 12 RD&D activities, one to three first-of-a-kind demonstration projects on a commercial scale, financing aspects, regulatory framework and support to internationalization. It also identifies the parties interested in carrying out each RD&D activity and potential financing sources, taking advantage of the strong synergies between research performers, industrial players, national funding agencies, and the European Commission.

Investments under the 12 RD&D activities included in the implementation plan amount to €200 million. The first-of-a-kind demonstration projects require up to €1 billion in addition. Funding mainly originates from private and public sources at national level.

Finally, "technology improvement opportunities" have been identified in the context of the US SunShot initiative [24]. These have been pursued by the US Department of Energy (DOE) in order to increase performance and reduce costs and/or risks. See Table 5.10.

Table 5.10 Technology improvement opportunities in the US SunShot Initiative

Area	Technology improvement opportunities
Solar collectors	• Reflectivity and mirror cleaning • Alignment, focusing, and tracking • Spectrum-splitting and integrated PV mirrors • Drives, structural support, and controls • Manufacturing and installation
Thermal receivers	• Solar-selective coatings • High-temperature receiver designs
Thermal energy storage	• Thermochemical energy storage • Phase change materials (PCMs) • Solid particles • Molten salts • Solid-state sensible storage
Power block	• SCO_2 Brayton cycles • Air Brayton Cycles • Combined cycle • Cooling
Soft costs	• Glares • Avian hazard

Source Mehos et al. [24]

References

1. Albrecht J et al (2015) Balancing demand-pull and supply-push measures to support renewable electricity in Europe. Renew Sustain Energy Rev 49:267–277
2. ALINNE (2015) Análisis del Potencial de Desarrollo de las Tecnologías Energéticas en España; Alianza por la Investigación y la Innovación Energética, CIEMAT. Available at http://www.alinne.es:8899/documents. Accessed Oct 2018
3. Blanco J (2018) CSP implementation plan. EERA JP-CSP 3rd general assembly meeting AEE INTEC (Graz, Austria). 16th Jan 2018. Available at http://www.solarconcentra.org/wp-content/uploads/2018/02/Implementation-Plan-CSP.PSA_.pdf. Accessed Nov 2018
4. Braun F et al (2011) Holding a candle to innovation in concentrating solar power technologies: a study drawing on patent data. Energy Policy 39:2441–2456
5. Cardoso J et al (2018) Towards and alignment of National research programmes and funding for CST technologies in Europe. American Institute of physicos conference proceedings 2018. Available at https://aip.scitation.org/doi/10.1063/1.5067132. Accessed Nov 2018
6. CSP Today (2014) Solar tower report 2014: cost, performance and thermal storage. CSP Today, London
7. Del Río P (2010) Analysing the interactions between renewable energy promotion and energy efficiency support schemes: the impact of different instruments and design elements. Energy Policy 38(9):4978–4989
8. Del Río P et al (2018) An overview of drivers and barriers to concentrated solar power in the European Union. Renew Sustain Energy Rev 81:1019–1029
9. EIB (2018) InnovFin–EU Finance for innovators. Available at http://www.eib.org/en/products/blending/innovfin/index.htm. Accessed Nov 2018
10. ESTELA (2012) Strategic research agenda 2020–2025. Estela, Brussels. Available at https://docs.google.com/file/d/0B2recxHj7XBvZDY4UGR6Z0hUUkk/edit. Accessed May 2018

11. European Commission (2017) The strategic energy technology (SET) plan. At the heart of energy research and innovation in Europe 2007–2017. Available at https://setis.ec.europa.eu/sites/default/files/setis%20reports/2017_set_plan_progress_report_0.pdf. Accessed Nov 2018
12. European Patent Office (2018) PATSTAT. Available at: https://www.epo.org/searching-for-patents/business.html Accessed Aug 2018
13. Griliches Z (1990) Patent statistics as economic indicators: a survey. J Econ Lit XXVIII:1661–1707
14. Groba F, Breitschopf B (2013) Impact of renewable energy policy and use on innovation a literature review. DIW discussion paper 1318. Available at https://www.diw.de/documents/publikationen/73/diw_01.c.426553.de/dp1318.pdf. Accessed June 2018
15. Hoogland O et al (2018) Study on impacts of EU actions supporting the development of renewable energy technologies. Technology sector report. Solar thermal energy (D 2.2.)—Draft for conference attendees. Trinomics for the EC (DG-RTD) Ref 2017/RDT/SC/IPP-05441-2017. Available at http://solarheateurope.eu/wp-content/uploads/2018/09. Accessed Nov 2018
16. IEA (2014) Technology roadmap solar thermal electricity. IEA, Paris. Available at www.iea.org. Accessed Aug 2016
17. IEA (2018) Energy technology RD&D overview. Available at https://www.iea.org/statistics/rdd/ Accessed Aug 2018
18. IRENA (2012) Renewable energy technologies: cost analysis series. Concentrating solar power, vol 1: Power sector. Issue 2/5. June 2012. Available at www.irena.org/documentdownloads/publications. Accessed Sept 2018
19. IRENA (2015a) Renewable energy technology innovation policy. Available at www.irena.org. Accessed May 2018
20. IRENA (2015b) RD&D for renewable energy technologies: Cooperation in Latin America and the Caribbean. Available at www.irena.org. Accessed May 2018
21. IRENA (2016) The power to change: solar and wind cost reduction potential to 2025. Available at www.irena.org/DocumentDownloads/Publications/IRENA_Power_to_Change_2016.pdf. Accessed Sept 2018
22. IRENA (2017) INSPIRE web platform. Available at http://resourceirena.irena.org/gateway/dashboard. Accessed Aug 2018
23. Islam MT et al (2018) A comprehensive review of state-of-the-art concentrating solar power (CSP) technologies: current status and research trends. Renew Sustain Energy Rev 91:987–1018
24. Mehos M et al (2016). On the path to SunShot: Advancing concentrating solar power technology, performance, and dispatchability. National Renewable Energy Laboratory, Golden. NREL/TP-5500-65688. Available at http://www.nrel.gov/docs/fy16osti/65688.pdf. Accessed Sept 2018
25. Mir-Artigues P, del Río P (2016) The economics and policy of solar PV generation. Springer, Berlin
26. Mowery D, Rosenberg N (1979) The influence of market demand upon innovation: a critical review of some recent empirical studies. Res Policy 8(2):102–153
27. Thompson Reuters (2008) Whitepaper using bibliometrics. A guide to evaluating research performance with citation data. Available at: http://ips.clarivate.com/m/pdfs/325133_thomson.pdf. Accessed Sept 2018
28. Trinomics (2018) Study on impacts of EU actions supporting the development of renewable energy technologies. Technology sector report. Study Commissioned by the EC
29. WIPO (World International Patent Organization) (2018a) Inside WIPO. What is WIPO? Available at: https://www.wipo.int/about-wipo/en. Accessed Aug 2018
30. WIPO (2018b). Patenscope. Available at https://patentscope.wipo.int/search/es/search.jsf. Accessed Aug 2018
31. Zarza E et al (2015) Future renewable energy costs: solar-thermal electricity. How technology innovation is anticipated to reduce the cost of energy from European solar-thermal electricity plants. KIC InnoEnergy Available at http://www.innoenergy.com/wp-content/uploads/2015/01/KIC_InnoEnergy_STE_anticipated_innovations_impact.pdf. Accessed Aug 2018

Chapter 6
Public Support Schemes for the Deployment of Plants

Several national and regional governments have implemented incentive programs to promote CSP deployment around the world. These public promotion schemes have been the main driver of CSP deployment. Although Spain and USA have been the main leaders, other countries have recently emerged and have a presence in the CSP market and still others can be expected to be so in the future. This section provides a discussion of the promotion policies which have been used to encourage the uptake of CSP from selected countries from around the world. These countries represent around 99% of total CSP capacity currently being deployed.

6.1 China

China is one of the largest countries in the world with 9.6 million squared Km and a population of 1.3 billion. China has experienced rapid economic and social development that has resulted in an average GDP growth of nearly 10 percent a year and has brought more than 800 million people out of poverty. It is the second largest economy in the world since 2010 [61] and could become the world's largest economy in 2030 [2].

Since 2008, China has been the largest single contributor to world growth. It is the largest trading nation in the world, the largest exporter, and the second largest importer of goods. Huge state investments in infrastructure and heavy industries and private sector expansion in light industries have been main contributors to China's growth [2].

However, according to China's current poverty standard, there were 55 million poor in rural areas in 2015 [61]. Social and environmental challenges include high inequality, rapid urbanization, an aging population, internal migration of labor, environmental pollution, and external imbalances. China's 13th Five-Year Plan (2016–2020) addresses those challenges through environmental and social measures.

P. Mir-Artigues et al., *The Economics and Policy of Concentrating Solar Power Generation*, Green Energy and Technology,
https://doi.org/10.1007/978-3-030-11938-6_6

6.1.1 Energy Situation

Electricity demand per capita has increased impressively in the last years. A threefold increase between 2003 (1379 kWh) and 2014 (3927 kWh) has been experienced [61]. According to this source, 100% of the population has access to electricity (both in urban and rural areas).

The Chinese electricity sector is still undergoing transformation toward a market-based system. Forty percent of generation is owned by the five largest generation groups, the rest by a multitude of small generators [61]. The electricity transmission grid is dominated by the State Grid Corporation and two smaller regional monopolies, whereas distribution is characterized by several thousand small players on provincial, prefectural, and county level [61].

Further reforms toward a liberalized electricity market were underway already in 2015 [54]. Regulators are piloting various market mechanisms including bilateral corporate PPAs, power exchanges, and spot markets to further liberalize the retail market [5]. China has some of the cheapest residential electricity in the world because it is subsidized by commercial and industrial rates. However, in March 2018, regulators announced a goal to reduce retail commercial and industrial tariffs by 10% [5].

On the supply side, the power mix in China as of 2016 was reliant on coal (68.2%), with hydro accounting for a 19% share. Wind (3.8%), nuclear (3.4%), gas (2.7%), PV (1.2%), biomass (1%), waste (0.2%), and oil (0.2%) account for the rest [26, 27]. Renewable energy capacity installed as of 2017 amounted to 618 GW. This represents a fourfold increase since 2008 (174GW). Hydro dominates RES installed capacity, with 312 GW, followed by wind (164 GW), PV (130 GW), and bioenergy (11.3 GW). CSP accounts for only 14 MW [33].

6.1.2 Promotion Policy

Over the last years, China has been trying to build an energy system which is cleaner, less carbon intensive, safer, and more efficient than the present one (Wang et al. [59]). Aligned with this objective, the 13th Five-Year Plan for national economic and social development states that the government will support green and clean production, promote the green transformation of traditional manufacturing, and establish low-carbon production and recycling in industry. Among other measures, the government has established a 15.4% renewable energy share target by 2020 and a 27.5% target by 2050.

China has a remarkable potential to develop CSP because of its abundant solar resource and available land area. In this regard, Wang et al. [59] indicate that the best locations are found in the western part of the country where DNI mean values are around 9 KWh/m^2 and the estimated available land amounts to 2.63 million km^2.

However, the potential CSP deployment is limited by two factors: (1) CSP requires very flat land, and (2) most power demand is in the eastern part of China, whereas the resource potential is mostly in the west.

CSP research activities in China started back in the 1970s and have been articulated under several programmes of the Ministry of Science and Technology, such as the National Basic Research Program (973 Program), the National High-tech R&D Program (863 Program), the National Science Foundation of China (NSFC), and the National Technical Innovation Fund for Medium- and Small-Size Enterprise. All give long-term support to CSP technology [15]. As a result, China is playing an important role in various CSP research fields, including the design and manufacture of condensers, collector fields, and control systems as well as heat exchangers and energy storage systems.

Furthermore, the so-called Golden Sun program aimed at boosting the solar sector was launched in 2009. Under this program, the central government supported half of the investment costs of large-scale solar power plants. As a result, in August 2012, the Beijing Badaling solar thermal power plant, funded by the National High-tech RD&D Program, started operation. Next, in 2014, the National Development and Reform Commission set a feed-in tariff of ¥1.2 per kWh[1] for the Delingha solar thermal power plant (50 MW).

As to the current and future prospects for CSP in China, the 13th Five-Year Plan foresees a 5 GW installed capacity target by 2020. The first batch of CSP demonstration projects consisting of 20 plants with a total capacity of 1.35 GW was issued by the National Energy Administration in September 2016. For this "pilot program," China announced a FIT of ¥1.15/kWh. The projects included tower, trough, and linear Fresnel technologies and most of the projects also incorporated storage. These projects had to be completely commissioned by 2018 in order to retain the FIT offer. According to ESTELA [15] and Wang et al. [59], such objective does not seem feasible given the current path of development and the above-mentioned limitations. As of today, the current prospects are that no more than 6 projects (out of the 20) of this first round will be commissioned before the end of 2018 and, in most cases, a deadline extension will have to be negotiated [38] (Table 6.1).

Beyond this 20-project pilot program, longer-term plans to deploy around 30 GW of CSP by 2030 have been announced. In response to this announcement, [15] has recently informed that some companies have already signed project development agreements with local governments and project sites have already been selected in high DNI western cities like Delingha, Hami, Yumeng, Akesai, and Goldmud.

In order to facilitate the development of a Chinese CSP industry which is able to accomplish such ambitious targets, efforts to improve the quality of the planning of CSP projects, establish a technology standard system, monitor experiences for demonstration projects, improve the economy and management of CSP projects, and develop a relevant electricity pricing policy to support CSP are currently in place (Wang et al. [59]). Furthermore, in order to promote technical innovation and build an industry technology innovation chain, the National Solar Thermal Energy

[1]¥1 = $0.13 (as of Dec 27, 2018).

Table 6.1 CSP projects in China

	Year operation	Technology	Capacity [MW]	Storage capacity [h]	LCOE [$/kWh]	Total cost [million $]	Remuneration [$/kWh]
Dacheng	2019	Fresnel	50	13	0.11	249	0.17
Delingha 50 MW	2018	Trough	50	9	0.12	287	0.17
Dunhuang 100 MW Phase II	2018	Tower	100	11	0.08	450	0.17
Gansu Akesai	2019	Trough	50	15	0.1	294	0.0
Golmud (2*100 MW)	2018	Tower	200	15	0.06	797	0.17
Hami 50 MW	2019	Tower	50	8	0.1	234	0.17
Qinghai Gonghe	2019	Tower	50	6	0.1	181	0.17
Rayspower Yumen 50 MW Thermal Oil Trough project	2019	Trough	50	7	0.0	222	0.17
Royal Tech Yumen	2020	Trough	50	9	0.1	199	0.17
SunCan Dunhuang 10 MW (I Phase)	2016	Tower	10	15	0.0	64	0.18
Supcon Phase I	2013	Tower	10	2.5	0.09	26	0.21
Supcon Solar	2018	Tower	50	6	0.1	156	0.18
Urad Middle Banner 100 MW Thermal Oil Parabolic Trough project	2019	Trough	100	10	0.1	415	0.17
Yumen Xinneng	2018	Tower	50	9	0.11	265	0.17

Source CSP guru [36]

Alliance was established in 2009, under the support of the Ministry of Science and Technology, the Ministry of Finance, the Ministry of Education, State-owned Assets Supervision and Administration Commission of the State Council, China Federation of Trade Unions, and China Development Bank. Up to now, the Alliance has 54 members in total, with the Institute of Electrical Engineering of the Chinese Academy of Sciences as the chairman member. The members involve companies of essential raw materials production, manufacturers of key equipment, system integration, and engineering construction as well as national renowned institutes and universities in optics, thermodynamics, mechanics, and materials [15].

As to the future prospects for CSP in China, policy support is expected to continue as CSP is not yet deemed a mature technology and the Chinese government has recognized its promising outlook. However, according to various authors (Wang et al. [59]) [58], some challenges will have to be addressed in order to succeed in this endeavor. First, scaling up CSP to a major energy vector will require large efforts both in terms of technology and cost development. Second, the geographical mismatch between the high solar resource in the west and the high power demand in the east will require special measures in terms of grid development when moving to large-scale CSP deployment. Finally, China will have to profit from stronger international collaboration in the field, standardization, and intellectual property rights legislation and management.

6.2 India

With a surface of 3.1 million square km, a current population of 1339 million people and a GDP of $2597 trillion (data for 2017, [62, 63]), India has experienced substantial increases in its social welfare levels. Its gross national income in 2017 was $1870 per capita, a level which has increased continuously since 2002 (from $450 per capita). Its GDP growth rate has been in the 6% to 8% range in the last years. Life expectancy at birth has reached 68 years, and poverty has declined since 2004 [62, 63]. Its population is expected to overtake China's in 2028 to become the world's most populous nation [2].

However, the country still faces considerable economic, environmental, and social challenges. Key issues confronting the Indian government include ensuring high growth levels, fostering faster job creation, addressing distress in the agricultural sector, and strengthening implementation of flagship government programs [62, 63]. According to recent estimates, 30% of the households are still below the poverty line [35]. India has constantly experienced a demand power supply gap, which is regarded as a major hindrance to its growth [4]. Capacity shortages, regular blackouts, structural underinvestment and market and institutional failures are current problems of the Indian power sector [32] (see below). In addition, high and increasing import dependence exposes the country to greater geopolitical risks and international price volatility.[2]

India has agreed to reduce the emissions intensity of GDP by 33–35% in 2030 (compared to 2005 levels) as part of the Intended Nationally Determined Contributions (INDCs) submitted to the United Nations Framework for Climate Change (UNFCCC) in preparation of the Paris Agreement. It has also committed to increase the share of RES capacity up to 40% of total generation capacity and to increase nuclear energy capacity from its current level of 5.7–63 GW by 2032 [32, 35, 47]. The Indian Government has also set up an ambitious target to build new renewable generation capacities of 175 GW by 2021–2022 [19].

6.2.1 Energy Situation

Although one-sixth of the world's population lives in India, it only accounts for 6% of global energy use and between 240 and 300 million people still lacks access to electricity [32, 35]. Its per capita electricity consumption is among the lowest in the world. A significant amount of demand is unmet, owing to limited availability and accessibility of electricity [35].

Electricity shortage problems have been caused by insufficient fuel supply and power generation and transmission capacity [4]. India's grid infrastructure is already strained and needs major improvements. Its transmission and distribution losses are among the highest in the world, averaging 26% of total electricity generation [32, 41]. According to Gupta [19], the distribution companies are unable to pay the generators due to huge transmission losses, there are inefficiencies in billing and collection, and there are political preferences and interferences in the distribution chain in the form of lower rates or waivers, resulting in huge losses and significant outstanding debts.

The existing Indian power system is divided into five regional grids (northern, southern, eastern, northeastern, and western). By the end of 2013, all these grids had been interconnected through high-voltage transmission lines, but interregional transmission capacity is relatively low compared to installed capacity [32]. While the electricity grid now covers much of the country, reaching rural or remote areas with the necessary transmission and distribution infrastructure often remains a challenge and, thus, there are supply constraints [32].

Grid infrastructure challenges partly stem from the high growth rates in electricity consumption by both industry and residents. The total length of transmission and distribution lines grew by just 50% in the 2002–2013 period, while total electricity demand doubled. Growing population and urbanization along with increasing electrification and per capita usage have driven and will drive growth in power con-

[2]Over 80% of the total oil requirement in India is imported.

sumption.[3] India's electricity demand has grown by 10% a year over the past decade. Peak power demand has been growing at an average of 4.5% over the last five years and is expected to accelerate in the future [5]. According to the Ministry of Power [41], the deficit in peak demand (difference between peak demand and peak demand being met) is 1.5 GW as of 2018, although it has been reduced in the last decade (15.1 GW in 2009 and 7 GW in 2014). It is expected that electricity consumption will reach around 1894 TWh in 2022 [24] (from 1216 TWh in 2016, according to IEA [26]) and between 5518 and 3740 TWh in 2047, depending on the scenario [35]. Therefore, massive investments in power generation capacity and related infrastructure will be required, creating an important opportunity for renewable energy deployment [32].

On the supply side, the power mix as of 2016 is reliant on coal (74.7%), although the other ¼ is relatively diversified with hydro (9.3%), gas (4.8%), wind (3%), bioenergy (3%), nuclear (2.5%), oil (1.5%), and PV (0.9%) [26]. Renewable energy capacity installed as of 2017 amounted to 106 GW (up from 45 GW in 2008 and 51 GW in 2010). Hydro dominates RES capacity, with 49.3 GW, followed by wind (32.8 GW), PV (19 GW), and bioenergy (9.5 GW). CSP accounts for only 229 MW [33]. The share of CSP is very small in the renewable energy mix of the country in spite of having large potential across the country. The economic potential of CSP is estimated at 571 GW at an annual DNI over 2000 kWh/m^2 and WPD $\geq$ 150 W/m^2 in India [47]. India was, in addition to China, the only other country in Asia with CSP capacity under construction by the end of 2017, with the 14 MW Dadri Integrated Solar Combined-Cycle plant expected to begin operation in 2018 [48].

The fossil fuel import dependency is high and rapidly increasing. 15% of coal demand is being met from other countries. The import dependency of crude oil is 76%. Domestic natural gas production is 1.3 EJ with 0.3 EJ being imported [32].

According to Climatescope [5], the power sector in India has been unbundled into distinct actors for generation, transmission, distribution, and retail, there are legally separate private companies at each segment of the power system preretail, there is an independent transmission system that dispatches according to market dynamics and is not susceptible to state interference, there are not significant barriers to private sector participation in generation, and there is a functioning competitive wholesale generation market. The power sector in India is managed by the Ministry of Power. Generation of power is handled by federal government-owned companies, state-level corporations, and private sector companies. The transmission of power is mainly handled by the Power Grid Corporation of India. The responsibility for distribution and supply of power lies with individual state distribution companies [5].

[3]In addition, power is not used efficiently since the electricity price does not cover production costs and the conversion efficiency of Indian power plants is quite low (the average gross efficiency of thermal power plants is about 30%) [32].

6.2.2 Promotion Policy

India's Electricity Act of 2003 paved the way for regulatory interventions, which supported and accelerated the deployment of renewable energy. It mandates state electricity regulatory commissions (SERCs) to fix quotas for the percentage of electricity being handled by the power utilities so as to procure power from renewable energy sources. It requires SERCs to determine the tariff for all renewable energy projects across their states [32].

A main boost to renewables came with the publication of the National Action Plan on Climate Change (NAPCC) in 2008, which suggested a new direction in India's climate policy. The NAPCC envisages renewable energy will represent around 15% of India's total final energy mix by 2020. To achieve this, a minimum share of renewable energy is pegged in the national grid at 5%, starting in 2009/10, and this is to be increased by 1% per annum [32].

The Government of India released its roadmap to achieve 175 GW capacity of renewable energy by 2022, which includes 100 GW of solar power and 60 GW of wind power [24]. This implies that renewable energy would then contribute close to 20% of the country's total power consumption. The government has raised the increase in solar power capacity from 22 to 100 GW. There are also targets of 10 GW for biomass and 5 GW from small hydropower, but there is not any target for CSP [32].

A Renewable Energy Purchase Obligation (RPO) has been introduced, requiring power distribution companies to buy 5–10% of their electricity from renewable sources or to purchase renewable energy certificates (RECs). Each REC (1 MWh) is tradable across the states, and there is a separate market for Solar Renewable Energy Certificates [32].

Within RES, solar energy is given special emphasis. In this context, the Jawaharlal Nehru National Solar Mission (JNNSM) is one of the eight National Missions under the NAPCC. The JNNSM was launched in January 2009 with a funding of $930 million [57] in order to diffuse solar power technologies across the country as quickly as possible in order to comply with the low-carbon pledges and improve energy security in a country with limited fossil fuel resources [16].

Structured in a period with three phases, the JNNSM aimed to deploy 20 GW of grid-connected and 2 GW of off-grid solar power until 2022. However, given the progress achieved, the solar power capacity target was increased in 2015 by five times in order to achieve 100 GW by 2022 (40 GW of rooftop solar and 60 GW of medium- and large-scale grid-connected power plants) [35]. The main features of the JNNSM policy to encourage CSP installations are listed in Table 6.2.

It is not clear what will be the respective shares of CSP and PV technologies in the aforementioned 60 GW [35]. The idea in the first phase of the JNNSM (2010–13) was to give equal emphasis to both PV and CSP technologies. Therefore, 500 MW was allocated to solar PV and another 500 MW to CSP technologies in Phase I. For CSP, seven projects (470 MW) were awarded, although only 225 MW had been implemented by the end of 2015 (see below). Three projects of 10 MW capacities each were additionally awarded through the so-called migration scheme of the Indian Ministry of New and Renewable Energy (MNRE). However, only 25% (2.5 MW) has been implemented [47]. Although the Phase I in JNNSM gave the same share for PV and CSP technologies, the share of CSP was reduced to 30% in Phase II, and states were asked to fulfill 60% of these targets [35].

To create demand and attract investment in RES in general and solar energy (PV and CSP) in particular, the government has provided various incentives, including feed-in tariffs, PPAs, generation-based incentives, a renewable purchase obligation, renewable energy certificates, a viability gap funding interest subsidy, and other benefits (tax benefits) [47: 650–651].

Finally, it should be mentioned that the policies at the state level in India play a crucial role in the deployment of renewable energy sources in general and solar and CSP in particular (see, e.g., Ummadisingu and Soni [57] for Rajastan and Pérez et al. [44] for Gujarat). State utilities are mandated to buy green energy via a long-term PPA from solar farms.

India has (partially) relied on auctions in order to promote solar energy in general and CSP deployment in particular. Indeed, it was one of the first countries to use auctions for this purpose [52]. Developers submit bids offering a certain discount to a reference tariff published by the Central Electricity Regulatory Commission. The PPAs are for fixed terms (25 years) without any escalation clause [35]. Each interested developer commit to build and commission the plant 28 months after signing the PPA [16]. As the price of solar was above market levels, it was packaged with cheap coal produced by public plants (the so-called bundling mechanism; see Table 6.2).

Table 6.2 Key features of CSP policies under JNNSM

Feature	**Objective**
Reverse auction/competitive bidding	To procure solar power in a cost-effective manner
Long-term PPA	To provide long-term revenue certainty to solar power generators all projects commissioned under Phase I could get a 25-year PPA
Guaranteed off-take	To provide off-take guarantee for the solar power generated Phase I projects was guaranteed off-take from NVVN (a government company)
Payment security scheme (PSS)	To provide partial payment security for solar project developers in case of a default by state distribution utilities. It ensures financial closure of projects sanctioned under Phase I
Bundling of power	To make the relatively expensive solar power affordable to distribution utilities. NVVN resells the solar power procured at a lower cost to distribution utilities after bundling it with government-owned cheaper coal power

Source Frisari and Stadelmann [16: 17]

The bundling mechanism helped made the expensive solar power more acceptable for the distribution utilities as the cost was reduced by 70% [20]. Differently from other countries (see below), higher tariffs are not provided for peak power or for dispatchability and, thus, the winning bids did not have a thermal storage component [16].

There have been three phases of the JNNSM at central government level: Phase I (2010–2013), Phase II (2013–2017), and Phase III (2017–2022). CSP was awarded only during Phase I (although also at state level). 500 MW of multiple projects at specified locations was auctioned in August 2010 in technology-specific auctions.

Bidders had to comply with some prequalification requirements: financial, technical, connection to the grid, water availability, and domestic content. The project developers needed to submit proof of land possession [12].

A prequalification requirement of 30% local content rule (LCR) was adopted. Bidders have to be a technology provider (or have a tie-up with one) with experience in design and engineering of CSP plants, or they should have achieved financial close for at least one project with that technology [44: 1871].

The Indian CSP auctions have led to low prices, but also to delays. In Phase I, 470 MW were awarded to CSP (seven plants). Most of these projects are parabolic trough (76%, 380 MW), followed by Fresnel (20%), solar tower (2%), and stirling (2%). However, as of 2017, only 229 MW had been built and entered into operation [32], although those 470 MW were scheduled to be commissioned by March 2013 [40]. Indeed, by 2014, only 50 MW were operational (the Godawari project) [52]. Many factors are cited as reasons for this delay: overestimation of DNI levels, a relatively recent technology, a weak supply chain, depreciation of the rupee, the drastic reduction in PV costs, too short lead times which prevented bidders from obtaining serious EPC offers in the short time given, too short deadlines to construct the projects, lenient qualification requirements for bidders in terms of CSP experience and financial strength, problems of grid connection, expensive financing leading to difficult financial closure and difficulties in securing land and water (see [12] for a full analysis). These delays were in spite of considerable penalties.

According to NREL [43] and CSP Guru [36], currently 10 projects in India are either operational or under construction (Table 6.3). An overwhelming majority of them are parabolic trough without storage. It is worth mentioning that there is a Fresnel plant (the Dhursar plant, operational since 2014). Most plants are relatively small ones.

Final prices paid in the JNNSM Phase I Batch I were between 175 and $204/MWh. This involves a considerable reduction from the ceiling price of $255/MWh from which bidders had to offer reductions. The low prices achieved in the Indian solar auctions have raised concerns regarding the timely completion as well as the economic viability of these projects. According to a report commissioned by the Australian government on India, the reverse auction in India carries with it considerable risk of adventurous bidders, leading to an allocation of projects among awarded bidders who may ultimately be unable to deliver [35]. Purohit and Purohit [47] recommend some design elements to be applied in CSP auctions in India in order to increase the chances that winning bidders actually implement their projects, including (a)

Table 6.3 CSP projects in India

Power Station	Year operational	Status	City	DNI [$kWh/m^2/year$]	Construction start	Technology	Capacity [MW]	Storage capacity (h)
Godawari	2013	Operating	Nokh	1667	2011	Trough	50	0
Dhursar	2014	Operating	Dhursar	1742	N. A.	Fresnel	125	0
Megha	2014	Operating	Khammam	1476	2011	Trough	50	0
Gujarat Solar One	2017	Operating	Kutch	1749	2015	Trough	25	9
Dadri ISCC Plant	2018	Construction	Dadri	N.A.	N.A.	N.A.	N.A.	N.A.
Abhijeet	N.A.	Construction	Phalodi	1985	2015	Trough	50	0
KVK Energy Solar Project	N.A.	Construction	Askandra	1940	2013	Trough	100	4
Diwakar	N.A.	Construction	Askandra	1938	2013	Trough	100	4
ACME Solar Tower	2011	Operating	Bikaner	N.A.	N.A	Tower	2.5	0
National Solar Thermal Power Facility	2012	Operating (demonstration)	Gurgaon	N.A.	N.A.	Trough	1	0

Source CSP guru [36] and NREL [43]

improving data on the available solar resource at the time of bidding, (b) allowance of minimum auxiliary/back up fuel and provision of minimum TES in the projects, (c) more realistic timelines for submitting bids and constructing plants, (d) stricter enforcement of penalties for missing deadlines, and (e) strengthening requirements for participating in the bidding."

6.3 Morocco

Morocco has a surface of 710,850 km^2 (including West Sahara), a GDP of $109,139 billion and a young and growing population of 35.7 million people (2017 data, [62]). Life expectancy is 75 years, and the poverty rate is 4.8%. GDP growth rates recovered in 2017 (4%), from the low rate in 2016 (1.2%), which in turn represented a sharp reduction with respect to the growth rate in 2015 (4.5%). The growth rate in 2018 is expected to be around 3% [62].

Unemployment is a major concern in Morocco. The unemployment rate was 10.2% in 2017, and it is especially prevalent among the young, the educated and women (26.5, 17.9, and 14.7%, respectively). The United Nations Human Development Index shows that Morocco faces marked socioeconomic inequalities, particularly between urban and rural areas [49].

6.3.1 Energy Situation

Electricity demand in 2016 reached 35.4 TWh, whereas electricity demand per capita was 997.6 KWh. It has grown at an annual average rate of 5.1% since 2010, slightly outpacing the average economic growth of 3.5%, due to the increase in population, access to electricity in rural areas, urbanization, and better standards of living [5]. Peak daily demand has also risen at a similar rate to total demand, by 3.2% in 2016 [49]. Electricity demand is expected to grow between threefold and fourfold in 2030 (i.e., between 95 and 133 TWh) [49]. According to Climatescope [5], meeting the high growth in electricity demand and the evening peak (currently with expensive diesel generators) remains a primary concern in securing new generating capacity.

As of 2016, the electricity generation mix is dominated by fossil fuel sources: coal (52%), followed by gas (18.3%) and oil (9%). Renewables have a non-negligible role, especially wind (9%) and hydro (5.1%). With 1.2%, CSP has a small share [26]. Renewable energy capacity installed as of 2017 amounted to 2.5 GW (up from 1.4 GW in 2008). Hydro dominates RES capacity, with 1.7 GW, followed by wind (1 GW). CSP accounts for only 180 MW, but more than the other solar technology (PV, 25 MW) [33].

Morocco lacks significant conventional energy sources of its own and imports about 90% of its energy supply. For the electricity sector only, oil purchases account for 24% of total imports, nearly 50% of the trade deficit and 10–12% of its GDP [49].

Some regulations have recently tried to decrease the dependence on foreign markets, phase out fuel subsidies by 2017, and increase the role of clean energy sources. The energy dependency ratio fell from 98% in 2008 to 93.3% in 2016 [49].

The Moroccan state utility Office National de l'Electricité et de l'Eau (ONEE) holds a monopoly for the transmission network (and its expansion) and is involved in all stages of the electricity supply chain [25]. It produces 1/3 of total electricity generation, buys electricity from Moroccan private producers, imports electricity, and is responsible for electricity distribution [5, 18]. It has the status of "single electricity buyer" except for RES generation, for which a specific law permits private-to-private power transactions [21]. However, in 2012, Morocco changed the national regulatory framework for the electricity sector, leading to a weaker dominant role of ONEE since it provides for a free market for the exchange of electricity from renewable sources among producers and customers [25].

Investment in the construction of power plants is generally made through public–private partnerships, which always involve one of the government energy agencies (usually ONEE or MASEN) [49]. The Moroccan Agency for Sustainable Energy (MASEN) is 25% owned by the state-owned ONEE, and it usually takes a 25% stake in private power projects [49]. It has been instrumental in reducing both barriers and risks in Moroccan renewables investment [5]. Although its task was only to implement the national solar power plan, it can also set up objectives for all renewable energy technologies now [18].

The electricity tariffs are not uniform. They vary according to the level of consumption, time of day, and type of meter [18]. The bihourly pricing, which is optional for users with consumption above 500 kWh, aims to reduce consumption during peak hours by encouraging discounted use during normal business hours [25]. In addition, ONEE does not pass on the full cost of generation to consumers and operates at a loss [5]. The rates are below the real average costs of production and transmission and represent a hidden subsidy to final consumers of around MDirham 0.30/kWh [25]. State subsidies granted to the energy sector accounted for 6% of the government budget (in 2013). However, these subsidies are expected to be gradually phased out [18].

6.3.2 Promotion Policy

Morocco has set two targets of 42% of RES in installed electricity capacity for 2020 (6 GW) and 52% (around 13 GW) for 2030. According to RES4MED [49], the 42% target is likely to be met.

Law 13–09 on renewable energy, enacted in 2010, established a legislative framework for the promotion of renewable energy investments, allowed the development of private energy production through IPPs, and granted investors the right to establish renewable energy projects, sell electricity directly to customers on the high-voltage market, and export unutilized energy [49]. It established the principle that any renewable power producer, both public and private, had the right to be connected to the medium-, high-, and very high-voltage national electricity grid (although not to the low tension network) [14]. Although power plants are generally built through public–private partnerships, investments can be wholly public or wholly private [18].

The Government of Morocco launched and financed the Morocco Solar Plan (MSP) in 2009 with the goal of developing 2 GW of solar power by 2020 in five selected locations (an annual production of around 18% of current electricity production). MASEN was set up in order to help develop the projects [44: 1871] and execute the MSP. The Moroccan government is tendering solar capacity to help meet such target. MASEN is responsible for the majority of utility-scale solar tenders in Morocco. It invites private developers to bid for the projects, supported by a 25-year fixed term PPA. Thus, the CSP projects are designed as a public–private partnership between MASEN and a private sponsor selected through the auction [49]. The 25-year PPA specifies two different prices (for base and peak load) to better remunerate the power that the plant will dispatch at times of peak electricity demand [16]. CSP policies under the Moroccan Solar Plan have four main features (Table 6.4).

Table 6.4 Main features of CSP under the Moroccan Solar Plan

Main features	Description
Two-stage competitive bidding	(See text)
Long-term PPA and guaranteed off-take	The contract obliges MASEN to guarantee the purchase of the agreed amount of solar power from the project
Private–public partnership (PPP)	The PPP model allows the government to share the costs and risks with international and private financiers and project developers. Performance guarantees issued by the private developer completely shield the domestic public actors from construction and technical risks
Guarantees for viability gap funding	To mitigate the off-taker default risk, the government has provided MASEN with a guarantee to ensure its financial viability. International financial institutions have awarded the government a credit facility to be used to cover MASEN financial obligations

Source Frisari and Stadelmann [16: 15]

CSP has been awarded in technology-specific, site-specific, and multi-round auctions.[4] Three rounds of auction have taken place between 2011 and 2016, with three projects being awarded contracts: NOOR I (160 MW, which started production in 2016), NOOR II (200 MW, it started production in 2018), and NOOR III (150 MW, expected to start production in October 2018). The NOOR I bidding process was launched in 2011 and awarded in 2013. NOOR II and III were launched in 2013 and awarded in 2015. The NOOR I and II projects are parabolic trough, whereas NOOR III is a solar tower.

There are seven CSP projects in Morocco, with different configurations (trough, tower and hybrid) (Table 6.5). The NOOR projects are relatively large size with storage. The difficulty and expense in meeting evening peak demand led to storage requirements being added to Moroccan solar tenders.

A main feature of the Moroccan CSP auctions is the active role played by MASEN, which takes an active role in conducting an initial environmental impact assessment for each site, in commissioning prefeasibility studies, providing financial backing to the bidders and securing concessional low-interest loans from international finance institutions [28]. MASEN managed most of the risks (financial, permits, grid access) and therefore lowered the risk premium of the project.

Another key design element in the CSP auction is the existence of local content. Under the MSP, bidders are encouraged to promote local manufacturing. For instance, in NOOR I, a 42% local content portion was included [14].

Regarding the results on winning prices, they have gone down from NOOR I ($0.189/kWh) to NOOR II ($0.14/kWh), although they increased for NOOR III ($0.15/kWh) [43]. The high cost of meeting the evening peak justified the relatively high PPA for NOOR I in 2013, which includes five hours of storage. Several factors are behind these low prices, but the aforementioned active role of MASEN is a main one (see [12] for a full analysis). Nevertheless, according to De Lovinfosse et al. [7], the auction process has been long and complex.

On December 31, 2015, MASEN launched a call for expression of interest for the development of the first phase of the NOOR Midelt solar project, which will comprise two plants (one will be PV and another will be solar thermal), both with storage and in the 150–190 MW range. The bidders may decide on the share of each technology that they use, but the facility must serve the evening demand peak [5].

[4]See del Río and Mir-Artigues [12] for a detailed analysis of the design and functioning of the auctions for CSP in Morocco.

Table 6.5 CSP projects in Morocco

Power Station	Year operational	Status	City	DNI [$kWh/m^2/year$]	Construction start	Technology	Capacity [MW]	Storage capacity [h]
Ain Beni Mathar	2011	Operating	Ain Beni Mathar	2072	2008	Hybrid	20	0
Noor I	2015	Operating	Ouarzazate	2497	2013	Trough	160	3
Noor III	2018	Construction	Ouarzazate	2508	2017	Tower	150	7
Noor II	2018	Operating	Ouarzazate	2503	2015	Trough	200	7
Ait-Baha Pilot Plant	2014	Operational (pilot)	Ait-Baha	N.A.	N.A.	Trough	3	5
eCare Solar Thermal Project	2014	Operational (demonstra-tion)	Undefined	N.A.	N.A.	Fresnel	1	2
IRESEN 1 MWe CSP-ORC		Under construction (pilot)	Ben Guerir	N.A.	N.A.	Fresnel	1	20 min

Source CSP guru [36] and NREL [43]

6.4 South Africa

With a surface of 1.22 million square km, a population of 56 million people and a GDP of 349,419 million $, South Africa, has one of the continent's biggest and most developed economies. However, it went into a deep economic crisis between 2011 and 2016, with a sharp reduction of GDP (from 416,878 billion $ in 2011 to 295,763 billion $ in 2016) and GNI per capita (from 7540$ in 2012 to 5430$ in 2017) [62].

Crucial economic and social challenges lie ahead for South Africa, including sluggish GDP per capita growth (between 0.5 and 1.3% since 2014), relatively high and recently increasing poverty levels (the poverty rate was 55% of population in 2014) and relatively low life expectancy at birth (62 years) [62]. High unemployment remains a key challenge, standing at 26.7% in 2017. South Africa has one of the highest inequality rates in the world (the Gini coefficient was 0.69 in 2014) [63].

Regarding climate policy, South Africa is a signatory to the UNFCCC Paris Agreement. Its INDC envisages an emissions increase in the range of 20% and 82% in 2030 above the emissions level in 1990, although Climatescope [5] estimates that the emissions trajectory as a result of the current policy framework will be above that range.

6.4.1 Energy Situation

South Africa has a semi-decentralized power distribution sector, with about 180 distribution companies. Unbundling of generation and transmission has not yet taken place. Eskom is a vertically integrated actor responsible for most of the power generation (95%), transmission, and distribution [9].

The Department of Energy (DOE) periodically sets how much new power generation is needed and from which sources (until 2030), based on the Integrated Resource Plan (IRP), released in 2010. The National Energy Regulator (NERSA) can only license new capacity within these limits [9].

Total electricity demand has increased from 205 TWh in 2000 to 225 in 2016 [26, 27] although it has been reduced in the last decade, from an all-time record of 238 TWh in 2007. Electricity consumption per capita also reached a peak in 2007 (4777 kWh) and then went down to 4198 kWh [62].

On the supply side, the power mix as of 2016 is reliant on coal (89%), followed by nuclear (6%), hydro (1.5%), wind (1.4%), PV (1%), CSP (0.2%), and biomass (0.1%) [26, 27]. Renewable energy capacity installed as of 2017 amounted to 4.9 GW (up from 0.8 GW in 2010). Hydro dominates RES capacity, with 3.4 GW, followed by wind (2.1 GW), PV (1.7 GW), and bioenergy (148 MW). CSP accounts for only 300 MW [33]. For the second year running, South Africa led the market in new additions in 2017, being the only country which brought new CSP capacity online [48].

6.4.2 Promotion Policy

In 2008, the Renewable Energy Feed-in Tariff (REFIT) program was introduced to encourage the participation of the private sector in electricity generation. In 2010, the IRP proposed that a capacity of 1200 MW CSP should be built in the 2010–2013 period. The IRP was revised in 2013 and proposed that CSP capacity should be increased from 1200 to 3300 MW.

The Department of Energy introduced the REIPPP in August 2011. It replaced the REFIT program, in order to implement the renewable energy allocations of the IRP [51]. The goal was to develop 3725 MW of renewable energy capacity by 2016, and the program was structured in bidding windows [44: 1871]. However, the scheme came to a halt in 2016, after Eskom refused to sign further PPAs until it received guidance from government [5].

In the first and second bidding rounds, the REIPPPP allocated a capacity of 200 MW to CSP. During the first bidding round, projects totaling up to 150 MW were awarded to CSP. The second bidding round awarded the remaining 50 MW to CSP.

During round three, a further 200 MW capacity was allocated to CSP. Therefore, multi-item, technology-specific, and multi-criteria auctions have taken place for CSP. The auctions were sealed-bid; PAB ones and the winners received a 20-year PPA indexed to inflation. The tariffs for the first and second bidding rounds were capped, based on the previously administratively set FITs (see [12] for further details). In 2013, the REIPPP program changed the tariff structure of the CSP to a two-tier tariff structure to encourage dispatchable CSP plants to deliver peak energy [51]. A multiplier of 270% of the base tariff applies to peak load hours, and no tariff applies to the night hours. In addition to material prequalification requirements (Table 6.6), bid bonds applied. These doubled once a bidder became a preferred bidder. For every day that the commercial operation date (COD) was delayed beyond its scheduled COD, the operating period of the contract would be reduced by an additional day [34]. Bidders were responsible for securing grid access.

In the second stage, bids were reviewed based on weighted criteria: 70% for their price offer and 30% for their additional contribution to economic development (i.e., over and above minimum requirements). Of the 30 points awarded for economic development, job creation counts for 25%, local content for 25%, ownership for 15%, management control for 5%, preferential procurement for 10%, enterprise development 5%, and socioeconomic development for 15% [9, 42].

An amount of 600 MW has been awarded in seven projects. Of these 600 MW, only 2/3 (400 MW) are operational as of May 2018. According to REN21 [48], several CSP projects under development faced ongoing uncertainty in 2017 as the state-owned utility (Eskom) delayed the signing of PPAs. However, progress was made in April 2018, when the Department of Energy signed 27 renewable project contracts with independent power producers, including one for a 100 MW CSP project.

Table 6.6 Prequalification criteria in South Africa's REIPPP

• Project structure: The bidder must provide a structural diagram showing its debt and equity participants, contractors, and key equipment suppliers
• Legal requirements: The bidder has to declare its acceptance of the terms of the PPA, the implementation agreement, and other designated project agreements
• Land acquisition and land-use requirements: Bidders must secure the project site and identify all permits and licenses regarding land rezoning, subdivision, and water use
• Environmental consent requirements
• Financial requirements: Price, method of financing, sufficient progress in securing financing, and proof of its ability to raise such financing
• Technical requirements: Information on the technology to be used, resource data, contractor capability and track record, and a cost estimate for the grid connection. CSP bidders had to demonstrate that their key contractors had experience in at least two projects of comparable scale [44: 1871]
• Economic development requirements: Share ownership by black South Africans and local communities, local content, job creation, preferential procurement, management control, socioeconomic development, and enterprise development
• Value for money: Projects must provide a net benefit to the South African government and consumer [39]

Source del Río [9: 9]

The average PPA (base) prices over successive rounds (indexed to April 2014) went down from 3.20 ZAR/kWh in BW1 to 3.00 ZAR/kWh in BW2, 1.74 ZAR/kWh in BW3, and 1.62 ZAR/kWh in BW3.5 [17]. However, Pérez et al. (2014: 1874) criticizes that there was a lack of competition in the first two rounds, resulting in very small discounts from the cap tariff and significantly higher prices than other plants being developed at the same time elsewhere. Project size limits (100 MW) -which discouraged economies of scale- stringent technological and other qualification requirements, burdensome administrative procedures, local economic development requirements and short time spans between the auction announcement and the deadlines for bid submissions may have also contributed to this outcome (see [12], for a detailed analysis).

Table 6.7 provides details on the CSP projects in South Africa. Parabolic trough dominates. It can be observed that projects are not larger than 100 MW, given the aforementioned size limits. All plants are equipped with storage.

6.5 Spain

Located in the Iberian Peninsula, with a population of 47 million and an area of 504,000 km^2, Spain is the fifth most populated and second largest country in the EU28. It is the Eurozone's fourth largest economy and one of the European countries which has been most affected by the recent economic crisis. Between 2009 and 2013,

Table 6.7 CSP projects in South Africa

Power station	Year operational	Status	City	DNI [kWh/m^2/year]	Construction start	Technology	Capacity [MW]	Storage capacity [h]
KaXu Solar One	2015	Operating	Pofadder	2963	2012	Trough	100	2.5
Khi Solar One	2016	Operating	Upington	2952	2012	Tower	50	2
Bokpoort	2016	Operating	Globershoop	2949	2013	Trough	50	9.3
Xina Solar One	2017	Operating	Pofadder	2960	2014	Trough	100	5.5
Kathu Solar Park	2018	Operating	Kathu	2830	2016	Trough	100	4.5
Ilanga 1	2018	Construction	Upington	2937	2015	Trough	100	4.5
Redstone Solar Thermal Power Plant	2018	Under development	Postmasburg	N.A.	N.A.	Tower	100	12

Source CSP guru [36] and NREL [43]

nominal GDP fell by an accumulated 6.7%, the unemployment rate rose to a record high of 25.7% in 2012, and household disposable income fell by 4.3% in nominal terms. Since then, Spain has started to emerge from the severe economic recession, posting four straight years of GDP growth above the EU average. Unemployment has fallen but remains high, especially among young people.

6.5.1 Energy Situation

In 2017, primary energy consumption was 128,084 ktep, which represented a 3.7% increase from 2016. The share of renewable was 12.2%, while oil accounted for 43%, natural gas for 21.4%, nuclear for 11.9%, and coal for 10.5% [1]. Compared to the average energy dependence of the 28 European member states (53%), Spain has a much higher energy dependence rate (77.4%). Thus, despite decreasing, such dependence on fossil fuels poses important environmental and energy security challenges. From 2016 to 2017, Spain increased its final energy consumption by 1.4%. Of the total final energy consumption, oil products represented 51%, electricity 23.4%, natural gas 16.5%, thermal renewable energies 6.4%, and coal 2.3% [1].

The Spanish electricity system has a large, diverse, and reliable power generation fleet. As of 2016, five sources accounted for a two-digit share: nuclear (21.3%), gas (19.2%), wind (17.8%), hydro (14.5%), and coal (13.6%). Oil (6%), biofuels (1.8%), PV (2.9%), CSP (2%), and waste (0.5%) account for the rest. As of 2017, renewable energy capacity installed in Spain amounted to 48 GW (up from 36 GW in 2008). Wind dominates RES capacity, with 23 GW, followed by hydro (16.7 GW), PV (4.9 GW), CSP (2.3 GW), and bioenergy (1 GW) [33].

Despite the high technological diversity, the electricity market structure is highly concentrated. There are five main power generation companies (Iberdrola, Gas Natural Fenosa, Endesa, EDP—Energias de Portugal, and E.ON) and four main electricity retailers in Spain. Market concentration, which had been falling for a number of years with the increase in smaller renewable energy generating companies, has remained stable in recent years.

The country has been a net exporter of electricity since 2004. Since that year, net exports have doubled, mainly owing to increasing exports to Portugal, which were made possible through growing interconnection capacity. Spain is a net exporter to Portugal and Morocco, and a net importer from France since 2011. The volume of Spain's cross-border electricity trade by country varies from year to year, mainly because of the weather (Table 6.8).

While the majority (95%) of CSP projects in Spain use parabolic trough technology, Spain also hosts projects using solar tower technology (one with molten salts and two with steam), Fresnel (two projects with a combined capacity of 31 MW), and a 22 MW hybrid parabolic trough biomass power plant. Around 40% of the CSP capacity has storage systems based on molten salt, which gives a lot of flexibility to the generation. The operation of some of these plants extends back to 2007, and their production has increased every year, with the plants meeting a greater share of

Table 6.8 CSP projects in Spain

Name	Start	Owner	Location (town)	Installed capacity	Technology	TES hours
Abengoa Solar	2007	Atlantica Yield	Sanlucar la Mayor	11.0	ST	1.0
Andasol-1 Central Termosolar Uno, SA	2008	Cubico Sustainable Investments Ltd	Aldeire-La Calahorra	50.0	PT	7.5
Abengoa Solar	2009	Atlantica Yield	Sanlucar la Mayor	20.0	ST	1.0
Andasol-2 Central Termosolar Dos, SA	2009	Cubico Sustainable Investments Limited	Aldeire -La Calahorra	50.0	PT	7.5
Extresol 1 S.L.	2009	Saeta Yield	Torre de Miguel Sesmero	50.0	PT	7.5
Acciona Energía S.A.	2009	Contour Global	Alvarado	50.0	PT	0
Iberdrola Energía Solar de Puertollano S.A	2009	Ence Energía y Celulosa	Puertollano	50.0	PT	0
Tubo Sol Murcia, S. A.	2009	Novatec	Calasparra	1.4	Fresnel	0
Manchasol 1 Central Termosolar Uno S.L.	2010	Cobra	Alcazar de San Juan	50.0	PT	7.5
Extresol 2 S.L.	2010	Saeta Yield	Torre de Miguel Sesmero	50.0	PT	7.5
Renovables Samca, S. A	2010	Renovables Samca	La Garrovilla	50.0	PT	7.5
Renovables Samca S A	2010	Renovables Samca	La Garrovilla	50.0	PT	7.5
Gemasolar 2006, S.A.	2010	Torresol Energy	Fuentes de Andalucía	19.9	ST	15.0
Acciona Energía S.A.	2010	Contour Global	Majadas de Tietar	50.0	PT	0
Acciona Energía S.A.	2010	Contour Global	Palma del Río	50.0	PT	0
Solnova Electricidad S.A.	2010	Atlantica Yield	Sanlucar la Mayor	50.0	PT	0

(continued)

Table 6.8 (continued)

Name	Start	Owner	Location (town)	Installed capacity	Technology	TES hours
Solnova Electricidad S.A.	2010	Atlantica Yield	Sanlucar la Mayor	50.0	PT	0
Solnova Electricidad S.A.	2010	Atlantica Yield	Sanlucar la Mayor	50.0	PT	0
Andasol-3 Central Termosolar Tres, SA	2011	SWM/Ferrostaal/Innogy/Rhein E/	Aldeire-La Calahorra	50.0	PT	8.0
Manchasol 2 Central Termosolar Dos S.L.	2011	Saeta Yield	Alcazar de San Juan	50.0	PT	7.5
Arcosol-50 S.A.	2011	Torresol Energy	San José del Valle	50.0	PT	7.5
Termesol-50 S.A.	2011	Torresol Energy	San José del Valle	50.0	PT	7.5
Acciona Energía S.A.	2011	Contour Global	Palma del Río	50.0	PT	0
Helioenergy Electricidad Uno, S.A.	2011	Atlantica Yield	Écija	50.0	PT	0
Solucía Renovables 1	2011	Plenium Partners	Lebrija	50.0	PT	0
Planta Termosolar Extremadura, S.L.	2012	NextEra	Navalvillar de Pela	50.0	PT	9.0
Extresol 3 S.L.	2012	Saeta Yield	Torre de Miguel Sesmero	50.0	PT	7.5
Africana Energía S.A.	2012	TSK/Magtel/Grupo Ortiz	Posadas	50.0	PT	7.5
Termosolar Borges, S.L.	2012	Abantia/Comsa/Generalitat de Catalunya	Les Borges Blanques	22.50	PT	0
Acciona Energía S.A.	2012	Contour Global	Orellana	50.0	PT	0

(continued)

Table 6.8 (continued)

Name	Start	Owner	Location (town)	Installed capacity	Technology	TES hours
Helioenergy Electricidad Dos, S.A.	2012	Atlantica Yield	Écija	50.0	PT	0
Helios I Hyperion Energy Investments, S.L.	2012	Atlantica Yield	Puerto Lápice	50.0	PT	0
Helios I Hyperion Energy Investments, S.L.	2012	Atlantica Yield	Puerto Lápice	50.0	PT	0
Solaben Electricidad 2 S.A.	2012	Atlantica Yield 70%/Itochu 30%	Logrosán	50.0	PT	0
Solaben Electricidad 3 S.A.	2012	Atlantica Yield 70%/Itochu 30%	Logrosán	50.0	PT	0
Solacor Electricidad 1 S.A.	2012	Atlantica Yield 87%/JGC Corporation 13%	El Carpio	50.0	PT	0
Solacor Electricidad 2 S.A.	2012	Atlantica Yield 87%/JGC Corporation 13%	El Carpio	50.0	PT	0
Aries Solar Termoeléctrica, S.L	2012	Celeo/Elecnor	Alcazar de San Juan	50.0	PT	0
Aries Solar Termoeléctrica, S.L	2012	Celeo/Elecnor	Alcazar de San Juan	50.0	PT	0
Astexol Extremadura 3, SL.	2012	Celeo/Elecnor	Olivenza	50.0	PT	0
Tubo Sol Murcia, S. A.	2012	EBL	Calasparra	30.0	Fresnel	0
Ibereólica Solar Morón S.L.U	2012	Grupo T-Solar	Morón de la Frontera	50.0	PT	0

(continued)

Table 6.8 (continued)

Name	Start	Owner	Location (town)	Installed capacity	Technology	TES hours
Ibereólica Solar Olivenza, S.L.U	2012	Grupo T-Solar	Olivenza	50.0	PT	0
Guzman Energía S.L.	2012	Plenium Partners/FCC/Mitsui	Palma del Río	50.0	PT	0
Planta Termosolar Extremadura, S.L.	2013	NextEra	Navalvillar de Pela	50.0	PT	9.0
Arenales	2013	OHL/Steag/REEFF	Morón de la Frontera	50.0	PT	7.0
Serrezuela Solar II, S. L.	2013	Saeta Yield	Talarrubias	50.0	PT	7.5
Solaben Electricidad 1 S.A.	2013	Atlantica Yield	Logrosán	50.0	PT	0
Solaben Electricidad 6 S.A.	2013	Atlantica Yield	Logrosán	50.0	PT	0
Enerstar Villena, SAU	2013	Plenium Partners/FCC/ Mitsui	Villena	50.0	PT	0

Source Protermosolar (as of Dec 2018)

demand as a result. The optimization of production and its perfect coupling to the power demand curve makes the value of CSP production particularly relevant among renewables. CSP plants cover 0.6% of the energy demand in Spain and employ a total of 5200 professionals [15].

6.5.2 *Promotion Policy*

Since the 1990s, RES-E has been supported through a feed-in tariff (FIT) scheme. The Spanish FIT for renewable energy is widely considered to have been quite effective in triggering RES-E and CSP capacity deployment. In 1998, the Royal Decree (RD) on the Special Regime (RD 2818/1998) gave renewable energy generators two options: (a) a fixed premium on top of the electricity market price or (b) a fixed total price (feed-in tariff) [11]. Renewable energy producers could sell their electricity to distributors or directly to the market. Successive modifications of the FIT scheme took place in 2004 (RD 436/2004) and 2007 (RD 661/2007). The 2007 modification, reflected in Royal Decree 661/2007, implemented a cap-and-floor system for the premium on top of the electricity market price.

CSP was first eligible for FIT support in RD 436/2004 [10]. The FIT of 0.27 €/KWh for CSP over 25 years for plants up to 50 MW in capacity enabled the rapid development of a leading CSP industry in a short period of time. By 2013, a total of 2.3 GW of CSP had been installed, employing a workforce of approximately 20,000 professionals [46].

However, in 2012, a moratorium on support for RES-E was passed. It had an abrupt effect on the rate of renewable energies deployment in Spain and led the CSP market to a complete standstill.

In January 2016, a new support scheme based on auctions was put in place, which implied an end to the renewable energy moratorium. The auctions were part of a regulatory framework set up in the Law 24/2013 of the electricity sector and developed in RD413/2014. This regulatory framework, which is quite complex and does not have an international precedent, is fully described in [8]. RES-E plants would receive the market price plus a "specific complementary remuneration," which has two elements, a remuneration for the investment and a remuneration for the operation of the plant. The remuneration for the investment (Rinv) refers to a payment per kW that allows installations to recover those investment costs which cannot be recovered by the sales of electricity in the market. This payment is received during the useful regulatory life of the installation. The remuneration for the operation (Ro) refers to a payment per kWh for those technologies whose operational costs are above the average wholesale electricity price.

Three auction rounds have taken place (January 2016, May 2017, and July 2017). As a result, 8700 MW of RES-E has been awarded contracts, although none for CSP (see Del Río [10] for further details).

As of today, the prospects and current outlook for CSP in Spain are presently not very optimistic due to the above-mentioned retroactive policy changes between 2012 and 2013 that canceled the feed-in tariff for existing plants retrospectively and replaced it with a lower revenue level, which eroded investors´ confidence in future projects. Furthermore, if the same auction design in Spain, with respect to technology neutrality, was to continue in the future, CSP plants would have low chances of being awarded. Despite the fact that CSP costs have declined drastically over the last few years, solar PV and wind are, as of today, much cheaper technologies [33]. CSP will be able to compete and be awarded only if the capacity of CSP to supply firm power was accounted for (i.e., by having an auction that requires the capacity to provide firm power).

According to consulted experts and industry representatives, it would be unfortunate that countries like Spain did not take advantage of the excellent conditions for the deployment of solar thermal power plants that would help the country move toward a largely renewable electricity generation, removing coal and nuclear power plants while declining system costs and boosting many sectors of the economy. As indicated by Luis Crespo in his inaugural speech at the CSP Today event held in Madrid in November 2018,[5] where he made reference to the latest report of Protermosolar [46], "energy efficiency and demand management, together with the adequate use of interruptibility contracts and the increase in interconnections, would be sufficient for the deployment of new solar thermal power, together with the technologies, wind, photovoltaics, and biomass, in addition to the contribution of hydraulics, the backing of gas can be limited at levels below 5% in 2030."

6.6 United Arab Emirates

With a GDP of $382 billion, a surface area of 83,600 km^2, and a total population of 9.4 million in 2017 (up from 3 million in 2000), the UAE is a high-income federal country made up of seven emirates (Abu Dhabi, Dubai, Ajman, Fujairah, Ras al Khaimah, Sharjah, and Umm al Quwain). It has a high GDP per capita ($41,197), which is among the ten highest in the world. However, its GDP growth rate has recently been sluggish (0.8% in 2017), although it is relatively high when considering the last 5 years (3.6% over the period 2013–2017). The country has a low unemployment rate (2.4% in 2017) and a high life expectancy (77 years) [62].

Each emirate has autonomy in local affairs (including energy). Over 85% of residents are expatriates. The country ranks 40th in the Human Development Index (2014). Although the government has tried to diversify the economy away from fossil fuels, the country still relies on its vast oil and natural gas resources [23]. CO_2 emissions per capita are relatively high (23 metric tons per capita), and total CO_2 emissions have increased from 51 million tons in 2000 to 191 million tons in 2016 [62].

[5] Madrid 13 and 14 of November 2018. Available at https://events.newenergyupdate.com/csp.

6.6.1 Energy Situation

Total electricity consumption has increased substantially from 38 TWh in 2000 to 120 TWh in 2016 [62]. Electricity consumption per capita in 2014 was 11,000 kWh and has been relatively stable since 1998.

Economic growth across the UAE has led to massive increases in the demand for electricity, which will more than double by 2020 [55]. Electricity demand in the UAE exhibits a strongly seasonal effect due to the changes in ambient temperature and humidity. The primary electricity loads in the UAE are cooling, lighting, refrigeration, and other appliance loads [30]. In turn, peak demand has substantially increased in the last few years. This trend will probably continue, given the UAE's economic and population growth [5].

On the supply side, the UAE is one of the world's largest hydrocarbon reserve holders and exporters. According to data from the Organization of the Petroleum Exporting Countries (OPEC), the UAE's proven oil reserves were the seventh largest in 2013, at 97 800 million barrels, and its gas reserves ranked sixth or seventh, at 6.1 trillion cubic meters. At 2.8 million barrels/day, the UAE is also the eighth largest oil producer, the world's third largest oil exporter, and the 17th largest gas producer [29].

However, the UAE is now facing its first-ever shortage of low-cost gas, As of 2016, gas accounted for 98.4% of electricity generation and oil for 1.2% [26].

With limitations on how much and how fast traditional energy resources, like natural gas, can be brought to market, as well as concerns about climate change, the UAE government launched various initiatives aimed at identifying alternative means for producing the power needed to fuel its economy [55]. However, the share of renewables is negligible (only 0.2% from CSP in 2016) [26]. As of 2017, a total 357 MW of accumulated capacity had been installed in the UAE, up from 14 MW in 2012 [33]. There is 1 MW of wind, 255 MW of PV, 100 MW of CSP, and 1 MW of bioenergy [33]. According to IRENA [29], the UAE's commercial case for renewable energy largely owes to avoidance of using high-cost natural gas for electricity production.

There are four main utilities in the UAE: ADWEA (Abu Dhabi), DEWA (Dubai), SEWA (Sharjah), and FEWA (Ajman, Fujairah, Ras Al Khaimah, Umm Al Quwain). Each has their own rate structures and planning and investment processes [29, 30]. They have a monopoly on procurement, transmission, and distribution. Generation is, however, open to private participation. In fact, over 90% of Abu Dhabi's generation is provided by IPPs [5].

Retail electricity tariffs are the Gulf's highest and are independently set by the four utilities [5]. Therefore, tariffs vary widely by emirate, are also differentiated between UAE nationals and expatriates, and are considerably higher than in other countries in the region. For example, in Abu Dhabi, tariffs have been raised to 21 fils[6]/kWh across all sectors, but UAE nationals are given a preferential rate of 5–5.5 fils/kWh [29].

[6]One hundred fils = 1 AED = 0.24.

Efforts are being made to interconnect transmission networks run by the four power authorities with a view to improving the grid's ability to cope with increased intermittency once solar generation begins ramping up [5].

6.6.2 Promotion Policy

The UAE has set itself the goal of reaching a 50% share of renewable energy in the generation mix by 2050. It announced its national clean energy target of 24% by 2021.

There is no national level target. Given the federal structure of the country, individual emirates have each determined their own objectives. The two largest emirates (Abu Dhabi and Dubai) have their own targets for renewables [31]. The target for Dubai is 7% of electricity generation coming from RES in 2020 and 15% in 2030, and the target for Abu Dhabi is 7% of electricity from RES in 2020 [48]. Dubai Clean Energy Strategy aims to generate 75 percent of Dubai's power from clean energy by 2050. Dubai recently increased its capacity target for the Mohamed Bin Rashid Al Maktoum Solar Park from 1 to 5 GW in total by 2030.

According to IRENA [29], the country has emerged as a significant investor in renewable energy globally and a political advocate for these technologies. However, Climatescope [5] argues that the country has not put in place any renewables incentives and that progress to date has largely been driven by state-backed projects. Investment in renewables has been considerable ($400 million for funding solar plants in 2017) [5].

In terms of renewable energy policy, the most relevant measure is competitive tendering for power plants. Projects are competitively tendered, with a tariff (originally at a premium to that for gas-fired generation) negotiated with the winner. The government typically retains a majority stake in the project, with independent power producers taking the remainder [29].

Solar power has been the primary focus of UAE efforts to date. Generally speaking, solar PV is increasingly seen as the most attractive technology in the UAE in the near term due to cost and resource availability. CSP with thermal energy storage, however, remains attractive for its potential to provide base load power [29]. The main challenges in deploying large-scale solar in the region are the dust particles/haze and humidity. A portfolio of solar PV and CSP with thermal energy storage installations provides good integration opportunities to meet daytime and evening demand requirements [29].

CSP has been promoted through auctions in the UAE, both in Abu Dhabi (Shams1) and Dubai (the Mohammed bin Rashid Al Maktoum Solar Park).

A CSP-specific, price-only, and site-specific auction was organized in Dubai in 2017. One 700 MW project was awarded. It has been designed to incorporate a 260-m solar tower and is expected to be the largest CSP facility in the world when completed. It will be built in a partnership between China-based Shanghai Electric and ACWA Power (REN 21 2018: 101). It combines a solar tower (100 MW) and three parabolic troughs (200 MW each). The project is required to generate power from 4 pm to 10 am. A 35 year PPA has been awarded, which is one of the main reasons behind the very low bid price of 7.3 cents$/kWh [37]. Other drivers of the low bid prices include: lenient prequalification requirements compared to other global tenders, a competitive local subcontractor market, low cost of finance (low commercial and regulatory risk), land provided at nominal costs, economies of scale (large project), long deadlines, and no local content requirements (see [12] for further details). However, the project is expected to be built in 2021, and, thus, it is too early to judge the effectiveness of the auction.

On the other hand, the Shams1 project was also awarded in an auction. The first invitation to bid for the construction of the CSP plant was made in 2008. Abengoa Solar, Total, ACWA Power, Iberdrola, Grupo Cobra, Grupo Sener, MAN Ferrostaal, and Solar Millennium participated in the bid. The bid selection process was delayed for two years. The design, build, and operate contract was finally awarded to Abengoa Solar and Total in June 2010 [45]. It is structured as an Independent Power Producer project, which is developed, owned, and operated by Shams Power Company PJSC, a joint venture between Masdar (60%), Total (20%), and Abengoa Solar (20%) [6]. Information about the bid prices and the 25-year PPA is not publicly available. The tariff was agreed between the consortium and Abu Dhabi's utility (ADWEA) and approved by the Regulation and Supervision Bureau [29].

Shams 1 has 100 MW of capacity, parabolic trough, and no thermal storage. It is already operating (since 2013). The $600 million, 2.5 km^2 plant has the capacity to feed power to 20,000 homes and divert 175,000 tons of CO_2 per year from the atmosphere [56]. It has attracted a 22-year, $600 million bank loan from eight foreign and two local banks led by the French bank BNP Paribas. The Shams consortium also received a million $153 equity subscription from its sponsors [53]. Furthermore, the plant earns carbon credits under the UN's Clean Development Mechanism [45].

Shams 1 is located in the middle of the desert and the plant faces atmospheric challenges like the high dust concentration, wind storms, and high ambient temperature [6]. However, despite these challenges, the plant has exceeded production expectations in the first five years of operations. It achieved a 99.5% reliability rate in 2016 and generated a total of 232 GWh, about 9% above expected output levels [22].

Table 6.9 provides comparative details on the CSP projects in the UAE.

6.7 USA

With a total surface of 9833,516 km^2 and a population of 316 million, the USA is the third largest country in the world. Due to its size, the USA encompasses a wide variety of climates and landscapes. It is the world´s foremost economic and military power. Its GDP is around a quarter of the world total [2]. As with many countries around the world, the US economy was hardly hit by the financial crisis and is currently recovering thanks to the strong fundamentals of the economy and

Table 6.9 CSP projects in UAE

Power station	Year operational	Status	City	DNI [$kWh/m^2/year$]	Construction start	Technology	Capacity [MW]	Storage capacity [h]
Shams 1	2013	Operating	Madinat Zayed (Abu Dhabi)	2019	2010	Trough	100	0
DEWA CSP Tower Project	Expected in 2020	Under development	Dubai	N.A.	N.A.	Tower	100	15
DEWA CSP Trough Project	Expected in 2020	Under development	Dubai	N.A.	N.A	Trough	600	10

Source CSP guru [36] and NREL [43]

pro-growth policies, market-based practices, and low regulatory environment that encourages flexibility and adaptation [50].

The political system of the USA is a federal republic that grants each of the 50 states the ability to enact many of their own laws and regulations [2].

6.7.1 Energy Situation

According to the IEA [27], the US natural gas boom has led to stable wholesale electricity prices, lower greenhouse gas emissions, and greater system flexibility. Meeting demand growth forecasts, however, requires significant investments in the US electricity system.

Electricity consumption per capita, which was 12,984 kWh in 2014, has remained more or less constant in the last 20 years. Non-renewable energy sources account for more than 4/5 of electricity generation in the USA. Gas (32.8%), coal (31.3%), and nuclear (19.4%) dominate the power mix, with hydro (6.8%), wind (5.3%), bioenergy (1.4%), PV (1.1%), oil (0.8%), waste (0.4%), and CSP (0.1%) accounting for the rest. Renewable energy capacity installed as of 2017 in the USA amounted to 230 GW (up from 116 GW in 2008). Wind dominates RES installed capacity, with 87 GW, followed by hydro (83.8 GW), PV (42 GW), and bioenergy (13.1 GW). CSP accounts for a non-negligible 1.7 GW [33].

6.7.2 Promotion Policy

Since the first CSP plants in the late 80 s, three primary incentives have been responsible for the growth of the CSP market in the USA [60]:

- The Federal Investment Tax Credit (ITC): The Energy Policy Act of 2005 created a 30% ITC for commercial and residential solar energy systems that applies to CSP. This system provides credits equal to 30% of the eligible property that was placed in service by the end of 2016.
- State Renewable Portfolio Standards (RPS): Most US states have now established a RPS, which requires an increased production of electricity from RES, such as wind and solar.
- Federal loan guarantees: The Department of Energy (DOE) is authorized to provide loan guarantees for projects that "avoid, reduce, or sequester air pollutants or greenhouse gases, employ new or significantly improved technologies, and provide a reasonable prospect of repayment."

As shown in Table 6.10, the SEGS I–VIII parabolic trough plants in California were the first to be deployed back in the 80s and early 90s and they benefited from the above-mentioned support measures. In total, the eight SEGS plants added up to 400 MW of installed capacity.

Table 6.10 CSP projects in USA

Name	Year	Technology	Capacity [MW]	Storage capacity [h]	LCOE [$/kWh]	Total cost [million $]	Remuneration [$/kWh]
Crescent dunes	2015	Tower	110	10	0.17	1015	0.14
Genesis	2014	Trough	250	0	0.19	1271	0.17
Ivanpah	2014	Tower	377	0	0.18	2300	0.14
Martin	2010	Hybrid; Trough	75	0	0.3	534	0.0
Mojave Solar Project	2014	Trough	250	0	0.24	1673	0.16
Nevada Solar One	2007	Trough	72	0.5	0.2	310	0.16
SEGS I	1984	Trough	13.8	3	0.66	127	0.0
SEGS II	1985	Trough	30	0	0.71	268	0.0
SEGS III	1985	Trough	30	0	0.25	202	0.0
SEGS IV	1989	Trough	30	0	0.23	180	0.0
SEGS IX	1990	Trough	80	0	0.27	391	0.0
SEGS V	1989	Trough	30	0	0.21	180	0.0
SEGS VI	1989	Trough	30	0	0.23	180	0.0
SEGS VII	1989	Trough	30	0	0.24	180	0.0
SEGS VIII	1989	Trough	80	0	0.25	406	0.0
Solana	2013	Trough	250	6	0.19	2124	0.15

Source CSP guru [36]

As the incentives ceased in the early 90s, CSP developments came to a standstill and did not resume until the construction of the Nevada Solar One plant that completed its construction in 2007. The next plant was the 75 MW Martin Project in 2010 in Florida.

In 2011, the American Recovery and Reinvestment Act (ARRA) was approved with the aim to support, with $6 billion, loan guarantees for renewable energy projects that included CSP technologies. Furthermore, the implementation of economic instruments such as the Modified Accelerated Cost Recovery System (MACRS) allowed solar installations to be depreciated over a period of 6 years. This measure accelerated the return on investment and reduced the tax liability in the first years of the plant operation, increasing the attractiveness for private investors [38].

At the state level, it is remarkable that the California Energy Commission and the Public Utilities Commission adopted the time-of-delivery (TOD) factors to estimate the electricity market price. The TOD factors take into account the varying energy and capacity values of electricity delivered in different times of the day. Due to its dispatchability, CSP can benefit from the TOD factors.

Complementary to the above-mentioned deployment support measures, in 2011, the US Department of Energy (DOE) launched the "SunShot Initiative" with a goal to support research activities which aim to reduce the costs of solar technologies. The specific goal of the initiative was to reduce the costs of solar energy by 75 percent, making it cost competitive at large scale with other forms of energy without subsidies [13].

The combination of the above-mentioned support schemes allowed for another burst of CSP deployment in the USA from 2013 to 2015. During this period, some outstanding projects were developed. For example, the Ivanpah project, which was completed in 2014, became the world's largest CSP solar tower facility with 392 MW of installed capacity. In that same year, the world's largest parabolic trough plant (Solana), with a 280 MW installed capacity, started its operation. Also in 2014, other parabolic trough plants started to operate, such as the Mojave Solar Project and the Genesis Solar, with capacities of 280 and 250 MW, respectively. The most recently completed plant in the USA is the high-profile Crescent Dunes 100 MW molten salt tower system in Nevada [3, 38].

After 2014, the main support schemes for CSP projects were no longer available. Most states had reached their RPS goals, the loan guarantee program was no longer available to fund utility-scale CSP projects, and the long lead time required to build CSP plants made the ITC unavailable because the projects would have to be placed in service by the end of 2016 [60].

As a result of the combination of the above-mentioned support measures, the USA has become the second largest market in terms of total installed capacity with 1745 MW.

According to Wilkins [60] and Bolinger and Seel [3], the future of CSP may be equally conditioned by some hurdles and drivers. On the one hand, the significant decrease in PV costs represents a challenge for the future market uptake of CSP. On the other hand, there are some factors on the horizon which bring some positivity. First, in 2014, the Environmental Protection Agency proposed a Clean Power Plan that would require each state to reduce CO_2 emissions from its utilities by about 30% by 2030. Second, the governor of California has recently proposed raising the state RPS to 50%. If implemented, this may be particularly important for CSP because California is a large consumer of electricity and has an excellent solar resource. Third, there is a possibility that the 30% ITC will be extended. Fourth, the increase of variable power in the electric grid puts a higher value of CSP plants′ availability to dispatch power whenever needed.

References

1. APPA (2018) Estudio del impacto macroeconómico de las Energías Renovables en España. Madrid
2. BBC (2018) Country profiles. Available at http://news.bbc.co.uk/2/hi/europe/country_profiles/default.stm. Accessed November 2018. Accessed Dec 2018
3. Bolinger M, Seel J (2018) Utility-scale solar: empirical trends in project technology, cost, performance, and PPA pricing in the United States. LBNL. Available at http://eta-publications.lbl.gov/sites/default/files/lbnl_utility_scale_solar_2018_edition_report.pdf. Accessed Nov 2018
4. Chakrabarti A (2015) Overview of energy sector in India. Energetica India 15:4–7
5. Climatescope (2018) Country reports. Available at http://global-climatescope.org/results/. Accessed Nov 2018
6. Crespo L (2017) Unlocking Sunbelt countries' solar potential. ESTELA. Available at http://www.estelasolar.org/Docs/WS/WS_17-01-12/ESTELA_Workshop_2017-01-2_LCR_Emirates.pdf. Accessed Nov 2018
7. De Lovinfosse I et al (2013): Lessons for the tendering system for renewable electricity in South Africa from international experience in Brazil, Morocco and Peru. Commissioned by: Deutsche Gesellschaft für Internationale Zusammenarbeit (GIZ) GmbH. South African—German Energy Programme (SAGEN). Department of Energy (DoE). (2013). Background Information on the REIPP Programme and the RFP. Available at www.ecofys.com/files. Accessed Nov 2018
8. Del Río P et al (2015) The new renewable electricity support scheme in Spain: a comment. Renew Energy Law Policy 6(1):17–24
9. Del Río P (2016) Report D7.1 implementation of auctions for renewable energy support in Spain: a case study of the Aures project. Available at www.auresproject.eu. Accessed 24 Sept 2018
10. Del Río P (2017) Status for implementation of auctions in Spain. Aures Project report. Available at http://auresproject.eu/news/status-implementation-of-auctions-in-spain. Accessed Sept 2018
11. Del Río P, Gual MA (2007) An integrated assessment of the feed-in tariff system in Spain. Energy Policy 35:994–1012
12. Del Río P, Mir-Artigues P (2019) Designing auctions for concentrating solar power. Energy Sustain Dev 48:67–81
13. DOE (US Department of Energy) (2012) SunShot Vision Study. Available at https://www.energy.gov/sites/prod/files/2014/01/f7/47927.pdf. Accessed Nov 2018
14. EIB (2015) EIB-FEMIP evaluating renewable energy manufacturing potential in the Mediterranean partner countries final report. Available at http://www.eib.org/attachments/femip_study_evaluating_renewable_energy_manufacturing_potential_en.pdf. Accessed Nov 2018
15. ESTELA (2018). Solar thermal electricity global market update—market situation by region—2018 Q1
16. Frisari G, Stadelmann M (2015) De-risking concentrated solar power in emerging markets: The role of policies and international finance institutions. Energy Policy 82:12–22
17. Gauché P et al (2017) System value and progress of CSP. Solar Energy 152, 106–139. Available at https://doi.org/10.1016/j.solener.2017.03.072. Accessed May 2018
18. GIZ (2017) Renewable energy and energy efficiency in Morocco Context and market access Information for small and medium-sized enterprises. Moroccan-German Energy Partnership PAREMA
19. Gupta A (2017) India's energy sector: Back in the game. Available at https://www.ey.com/Publication/vwLUAssets/ey-indias-energy-sector/$File/ey-indias-energy-sector.pdf. Accessed Nov 2018
20. Haas J et al (2018) Sunset or sunrise? Understanding the barriers and options for the massive deployment of solar technologies in Chile. Energy Policy 112:399–414
21. Hamane T (2016) A Snapshot of Morocco's Power Sector. 2016 Africa Energy Yearbook. Available at http://www.energynet.co.uk/webfm_send/2025. Accessed Nov 2018

22. Hashem H (2017) Dubai CSP developers urged to apply Shams 1 desert findings. New Energy Update. Available at http://analysis.newenergyupdate.com/csp-today/dubai-csp-developers-urged-apply-shams-1-desert-findings. Accessed Nov 2018
23. Howari F, Nazzal Y (2017) On The future of energy and resource security in the United Arab Emirates.Zayed University. Working Paper No. ZU-WP 2017-04-01. Available at https://www.zu.ac.ae/main/en/research/publications/_documents/On%20The%20future%20of%20energy%20and%20resource%20security%20in%20the%20United%20Arab%20Emirates.pdf. Accessed Nov 2018
24. IBEF (2018) Power Sector in India Available at https://www.ibef.org/industry/power-sector-india.aspx. Accessed Nov 2018
25. IEA (2014) Morocco 2014. International Energy Agency. Available at https://www.iea.org/publications/freepublications/publication/Morocco2014.pdf. Accessed Nov 2018
26. IEA (2018a). Global energy data statistics. International Energy Agency. Available at https://www.iea.org/statistics. Accessed Dec 2018
27. IEA (2018b). Country statistics. International Energy Agency. Available at https://www.iea.org/countries/. Accessed Dec 2018
28. IRENA (2013) Renewable Energy Auctions in Developing Countries. Abu Dhabi, Masdar Institute/IRENA. Available at https://www.irena.org/documentdownloads/publications/irena_renewable_energy_auctions_in_developing_countries.pdf. Accessed Nov 2018
29. IRENA (2015a) Renewable Energy Auctions: A Guide to Design. Abu Dhabi, Masdar Institute/IRENA. Available at http://www.irena.org/publications/2015/Jun/Renewable-Energy-Auctions-A-Guide-to-Design. Accessed Nov 2018
30. IRENA (2015b) Renewable Energy Prospects: United Arab Emirates, REmap 2030 analysis. Abu Dhabi. Masdar Institute/IRENA. Available at www.irena.org/remap. Accessed Nov 2018
31. IRENA (2016) Renewable Energy in the Arab Region. Overview of developments. Abu Dhabi, Masdar Institute/IRENA. Available at https://www.irena.org/-/media/Files/IRENA/Agency/Publication/2016/IRENA_Arab_Region_Overview_2016.pdf. Accessed Nov 2018
32. IRENA (2017) REMAP Renewable Energy Prospects for India. Working Paper. Abu Dhabi, Masdar Institute/IRENA. Available at https://www.irena.org/-/media/Files/IRENA/Agency/Publication/2017/May/IRENA_REmap_India_paper_2017.pdf. Accessed Nov 2018
33. IRENA (2018) Renewable Capacity Statistics. Available at https://www.irena.org/publications/2018/Mar/Renewable-Capacity-Statistics-2018. Accessed Nov 2018
34. Kruger W, Eberhard A (2018) Renewable energy auctions in sub-Saharan Africa: Comparing the South African, Ugandan, and Zambian Programs. WIREs Energy Environ 95:1–13
35. Kumar M, Ghosh S (2017) 100 GW solar power in India by 2022 – A critical review. Renew Sustain Energy Rev 73:1041–1050
36. Lilliestam J et al (2018) CSP Guru. New version of the data set including Chinese projects and additional technical, economic, financial, and industrial data. Available at. https://csp.guru/. Accesed May 2018
37. Lilliestam J, Pitz-Paal R (2018) Concentrating solar power for less than USD 0.07 per kWh: finally the breakthrough? Renewable Energy Focus 26:17–21
38. Lovegrave K et al (2018) Concentrating Solar Thermal Power Technology Status. Australia Renewable Energy Association, AIT P Thermal P Ltd. Forthcomming
39. McFarlanes (2015) South Africa's Renewable Energy Programme. An African success story. London, Clean Energy Africa Finance Guide: 2015 Edition
40. MNRE (2012). Jawaharlal Nehru National Solar Mission Phase II – Policy Document. Ministry of New & Renewable Energy. Available at https://mnre.gov.in/file-manager/UserFiles/draft-jnnsmpd-2.pdf. Accessed Nov 2018
41. Ministry of Power (2018) Power Sector at a Glance All India. Available at https://powermin.nic.in/en/content/power-sector-glance-all-india. Accessed Nov 2018
42. Montmasson-Clair G, Ryan G (2014) Lessons from South Africa's Renewable energy regulatory and procurement experience. Journal of Economic and Financial Sciences 7(S): 507–526
43. NREL (2018) Concentrating solar power projects database. https://solarpaces.nrel.gov/ Accessed Nov 2018

44. Pérez I et al (2014) National incentive programs for CSP–Lessons learned. Energy Procedia 49:1869–1878
45. Power Technology (2014) Shams 1 Solar Power Station Project. Available at https://www.power-technology.com/projects/shamsonesolarpowerst/. Accessed Nov 2018
46. Protermosolar (2018). Informe de transición del sector eléctrico. Horizonte 2030. Otro mix es posible (y deseable). Available at https://www.protermosolar.com/informe-de-transicion. Accessed Nov 2018
47. Purohit I, Purohit P (2017) Technical and economic potential of concentrating solar thermal power generation in India. Renew Sustain Energy Rev 78:648–667
48. REN21 (2018) Global Status Report Available at http://www.ren21.net/status-of-renewables/global-status-report. Accessed Nov 2018
49. RES4MED (2018) Country profile Morocco. Renewable Energy Solutions for the Mediterranean. Available at https://www.res4med.org/wp-content/uploads/2018/06/Country-profile-Marocco-2.pdf. Accessed Nov 2018
50. SACC (2018) USA country profile. Swedish-American Chambers of Commerce of the USA. Available at https://www.sacc-usa.org/export-guides/doing-business-in-us/us-country-profile. Accessed Nov 2018
51. Silinga C et al (2015) The South African REIPPP two-tier CSP tariff: Implications for a proposed hybrid CSP peaking system. Energy Procedia 69:1431–1440
52. Stadelmann M et al (2014) The Role of Public Finance in CSP: Lessons Learned. San Giorgio Group Policy Brief/Climate Policy Initiative. Available at http://www.climatepolicyinitiative.org. Acessed Dec 2016
53. Standards and Poors (2011) Q&A: Shams 1 Solar Power Project Financing Shines A Light On Renewable Energy in the Gulf. Available at http://cebcmena.com/wp-content/uploads/2018/01/Standard-Poors-on-Shams-1-Solar-Power-Project.pdf. Accessed Nov 2018
54. Steinhilber S (2016) Onshore wind concession auctions in China: instruments and lessons learnt. Report D4.1-CN of the EU-funded AURES project, Mar 2016. Available at http://auresproject.eu/sites/aures.eu/files/media/countryreports/pdf_china.pdf. Accessed Dec 2018
55. UAE (2018a) Energy in the UAE. UAE embassy in Washington. Available at https://www.uae-embassy.org/about-uae/energy-uae. Accessed Nov 2018
56. UAE (2018b) Energy. UAE government. Available at https://government.ae/en/information-and-services/environment-and-energy/water-and-energy/energy. Accessed Nov 2018
57. Ummadisingu A, Soni M (2011) Concentrating solar power—technology, potential and policy in India. Sustain Renew Energy Rev 15:5169–5175
58. Vieira de Souza LE, Gilmanova Cavalcante AM (2017) Concentrated solar power deployment in emerging economies: the cases of China and Brazil. Renew Sustain Energy Rev 72:1094–1103
59. Wang J, Yang S, Jiang C, Zhang Y, Lund PD (2017) Status and future strategies for Concentrating Solar Power in China. Engergy Sci Eng 5(2):100–109
60. Wilkins F (2015) U.S. market summary for an update to the Greenpeace 2009 Report on CSP
61. World Bank Group (2014) Performance of renewable energy auctions—experience in Brazil, China and India. Available at http://documents.worldbank.org/curated/en/842071468020372456/Performance-of-renewable-energy-auctions-experience-in-Brazil-China-and-India. Accessed Dec 2018
62. World Bank (2018a) Country data. Available at https://data.worldbank.org/country. Accessed Nov 2018
63. World Bank (2018b) The World Bank in South Africa. Overview. Available at https://www.worldbank.org/en/country/southafrica/overview. Accessed Nov 2018

Chapter 7
Summing Up

CSP is a dispatchable renewable electricity technology which might contribute substantially to a sustainable energy transition everywhere, in tandem with an increasing penetration of variable renewable energy technologies. According to the IEA [8], it could represent as much as 11% of electricity generation in 2050, with 954 GW of installed capacity (up from 5 GW today). CSP has a main distinguishing feature compared to other renewable energy technologies: It is able to provide dispatchable electricity, which allows balancing intermittent renewable electricity sources if these achieve a high penetration in the future. CSP plants contribute to grid balancing, spinning reserve, and ancillary services. They can also shift generation to when the sun is not shining and/or maximize generation at peak demand times [14]. However, with a current share in electricity generation worldwide of 0.1% [12], the technology currently plays a minor role in the power mixes everywhere.

With respect to the rest of renewable energy technologies, the future prospects of CSP generation depend on several factors: the evolution of the generation costs of CSP and the alternative renewable energy technologies (which, in turn, is highly influenced by capital costs and, to a lesser extent, the annual expenditures on O&M), grid infrastructure investment requirements for the transmission of the electricity from the production to the consumption areas, the evolution of the costs of storing electricity at large scale, the valuation and demand of dispatchability in electricity systems and/or in policy support frameworks.[1] To start with, the most pertinent comparison is with the cost of PV generation [2, 4: 378–386 and 1004, 6: 831 and 841, 9: 17]. A few years ago, both PV and CSP required considerable regulatory support in order to ensure their economic feasibility [10]. Furthermore, the cost of the PV kWh was twice the cost of the CSP kWh. However, after the drastic reduction of the price of panels between 2008 and 2012, the situation has totally reversed. In the middle of the last decade, the cost of PV was above \$0.3/kWh [11], whereas the cost of CSP was around \$0.2/kWh (see Chap. 3). However, in 2016, some PPAs for CSP were signed

[1] Another important use of CSP to be considered is desalination.

P. Mir-Artigues et al., *The Economics and Policy of Concentrating Solar Power Generation*, Green Energy and Technology,
https://doi.org/10.1007/978-3-030-11938-6_7

at $0.12/kWh. An auction in Dubai in 2017 (led by the Dubai Electricity and Water Authority) awarded CSP capacity at $0.073/kWh, and an auction in South Australia in that same year awarded capacity at $0.06/kWh for a 150 MW solar tower to be erected near Port Augusta [7, 9].[2] It should be taken into account that those locations have good conditions in terms of DNI (more so in the case of South Australia) and access to good harbor and road infrastructures, which facilitate the transport of equipment to the plant location (not far from the consumption areas). When it comes to PV, one of the last 2017 tenders in Germany—which is not precisely a sunny country—resulted in prices for utility-scale photovoltaic that ranged between €0.0429/kWh ($0.0505/kWh) and €0.0506/kWh ($0.0595/kWh). In early 2018, submitted bids reached an average value of €0.0433/kWh ($0.0527/kWh), with the lowest bid of €0.0386/kWh ($0.047/kWh) [5].[3]

On the other hand, advances in the design and cost reduction of batteries and other electricity storage options erode the strongest comparative advantage of CSP with respect to variable renewable sources: its dispatchability. Authors like [3] envisage considerable reductions in the price of lithium-ion batteries, both for electric cars and static storage: from $1000/kWh in 2010 to $162/kWh in 2017, and a prediction of $74/kWh for 2030. Taking these figures into consideration, the learning ratio, or the price decrease for every doubling of productive capacity, could be set at 19% [9].

Taking into account the above-mentioned considerations and assuming that there will be a need for flexibility (particularly with higher shares of variable renewables), the future widespread deployment of CSP may be contingent upon (1) the level of transmission costs and (2) the cost of other electricity storage solutions. Therefore, the condition for the widespread diffusion of CSP generation (MWh) can be expressed with the following conceptual formula:

Generation costs CSP+TES	$\oplus$	Required grid infrastructure cost	$<$	Generation costs of alternative technologies	$\oplus$	Cost of electricity storage

The key question is: will the expected reduction in the generation costs of CSP + TES plants exceed the reduction in the cost of competing renewable energy technologies, taking into account the evolution of the electricity transmission costs and the costs of storing electricity at large scale? As it is well known, the combination of generation and heat storage has the attractiveness of dispatchability.[4] This feature

[2] Promotion of renewable electricity generation in the State of South Australia includes the indicated solar tower and, among other milestones, a 100-MW (129 MWh) battery. This is the largest in the world so far.

[3] It should be taken into account that the PV sector is a source of spectacular news. For example, [1] explain that some improvements in metal halide perovskites cells have increased their performance to 21.5%. This could lead to a cost of the kWh in a utility-scale PV plant of €cents ~1.5.

[4] The cost of hybridization should be added. However, it has been ignored for reasons of simplicity and because it seems plausible to assume the existence of CSP plants in the future with only solar operation all hours of the year.

of CSP, plus the reduction in the generation costs, competes with (onshore and offshore) wind and PV (large-scale plants) generation costs, plus the addition of the storage cost (which is also expected to decline). All in all, the Achilles heel of CSP generation may be the long distance that exists between most of the best electricity generation locations (in terms of solar resource) and the consumption areas, which leads to a requirement to invest in long and costly transmission lines. Although all renewable sources use local resources (solar irradiation, biomass, wind, etc.), the challenge for CSP generation is that, in general, the better locations in terms of DNI are not those where the population is concentrated, with the exception of a few countries whose cities are mostly located in the desert, such as those in the Arabian Peninsula. In most cases, however, the places with the best generation conditions are hundreds or thousands of kilometers away from consumption centers, as it is the case in Australia, Chile, China, Mexico, or USA. This represents a much greater handicap if transmission grids have to cross international borders and gets even worse if there are important geographical barriers in the middle, as in Western and Central Europe since the Mediterranean Sea is located between the Sahara desert and the consumption centers. Although the impact of transmission cost on the competitiveness of CSP generation is a complex issue (since it all depends on how that cost is shared between generators and consumers), the need to build long transmission lines does not favor the prospects for CSP. This is clearly a disadvantage which, as it is also the case with wind offshore, hardens the competitiveness condition, which requires sharp reductions in the electricity generation costs.[5] However, this requirement is partly mitigated by the advantage provided by its dispatchability.

Although the magnitude of the future deployment of CSP globally remains uncertain due to the aforementioned factors, it is expected that CSP will be most deployed in those geographical regions with ideal conditions, that is, high value of direct normal irradiation, accessibility to water resources, flat terrain, and proximity to natural gas pipelines and the electricity transmission grid. In these places, it can also provide a relevant service: to cover the needs for desalinated water and electricity for urban, agricultural, and industrial uses. It is in these regions where CSP generation will better resist the pressure of competing technologies and, particularly, PV.

References

1. Abdi-Jalebi M et al (2018) Maximizing and stabilizing luminiscense from halide perovskites with potassium passivation. Nature 555(7697):497–501
2. Bensebaa F (2011) Solar based large scale power plants: what is the best option. Progr Photovolt: Res Appl 19:240–246
3. Curry C (2017) Lithium-ion Battery Costs and Market. Squeezed margins seek technology improvements and new business models. BNEF Reports, Bloomberg New Energy Finance July

[5] Maybe increased transmission cost could partially by offset by larger scope projects (which take advantage of economies of scale) far away from any urban and/or industrial area, that is, where there are no problems of space availability.

5, 2017. Available at https://data.bloomberglp.com/bnef/sites/14/2017/07/BNEF-Lithium-ion-battery-costs-and-market.pdf. Accessed Jan 2018
4. Edenhofer O et al (eds) (2012) Renewable energy sources and climate change mitigation. Special report of the intergovernmental panel on climate change. Cambridge, Cambridge University Press
5. Enkhardt S (2018) Germany's auction for large-scale solar: Bids below €0.04/kWh for the first time. February 20, 2018. Available at http://www.pv-magazine.com. Accessed Mar 2018
6. GEA (2012) Global energy assessment. Toward a sustainable future. Cambridge and Laxenburg, Cambridge University Press and IIASA
7. Hill JS (2018) South Australia approves world's largest single-tower thermal solar plant. Clean Technica News. 12 Jan 2018. Available at https://cleantechnica.com. Accessed Feb 2018
8. IEA (2014) Technology roadmap solar thermal electricity. International Energy Agency (IEA), Paris. Available at https://doi.org/10.1007/SpringerReference_7300. Accessed Dec 2018
9. IRENA (2018) Renewable power generation costs in 2017. International Renewable Energy Agency, Abu Dhabi. Available at http://www.irena.org/publications. Accessed Apr 2018
10. Madlener R, Mathar T (2010) Development trends and economics of innovative: solar power generation technologies: a comparative analysis. FCN working paper No. 1/2009. Institute for future energy consumer needs and behavior (FCN), FBE/E.ON ERC, Aachen University. Available at http://www.eonerc.rwth-aachen.de/fcn. Accessed Sep 2013
11. Mir-Artigues P, del Río P (2016) The economics and policy of solar photovoltaics generation. Springer, Switzerland
12. REN21 (2018) REN21: renewables 2018 global status report. Available at: http://www.ren21.net/status-of-renewables/global-status-report/. Accessed Jan 2019
13. Wang J et al (2017) Status and future strategies for concentrating solar power in China. Energy Sci Eng 5:100–109
14. World Energy Council (2016) World energy resources. Available at http://www.worldenergy.org/wp. Accessed Jan 2019